Through a Modern Microscope

Optical experiments on soft matter

Online at: https://doi.org/10.1088/978-0-7503-1431-2

Through a Modern Microscope

Optical experiments on soft matter

Mohan Srinivasarao
School of Materials Science and Engineering, Georgia Institute of Technology, Atlanta, GA, USA
and
School of Chemistry and Biochemistry, Georgia Institute of Technology, Atlanta, GA, USA

Jung Ok Park
Emeritus, School of Materials Science and Engineering, Georgia Institute of Technology, Atlanta, GA, USA

Karthik Nayani
Department of Chemical Engineering, University of Arkansas, Fayetteville, AR, USA

IOP Publishing, Bristol, UK

ISBN 978-0-7503-1431-2 (ebook)
ISBN 978-0-7503-1432-9 (print)
ISBN 978-0-7503-1913-3 (myPrint)
ISBN 978-0-7503-1433-6 (mobi)

DOI 10.1088/978-0-7503-1431-2

Version: 20251201

IOP ebooks

British Library Cataloguing-in-Publication Data: A catalogue record for this book is available from the British Library.

Published by IOP Publishing, wholly owned by The Institute of Physics, London

IOP Publishing, No.2 The Distillery, Glassfields, Avon Street, Bristol, BS2 0GR, UK

US Office: IOP Publishing, Inc., 190 North Independence Mall West, Suite 601, Philadelphia, PA 19106, USA

Contents

Preface

This text is the result of a laboratory course that was taught for about 2.5 years at The Georgia Institute of Technology. It was designed as a hands-on, open-ended, lab course that uses a compound light microscope, both in the transmitted and reflected modes, on a variety of materials that can be categorized as 'soft matter' or 'soft materials', to illustrate the principles of materials science and engineering that are often taught in introductory courses in a materials department. The idea of developing a laboratory course for students in the School of Materials Science and Engineering (MSE) occurred to us when two small departments, School of Polymer, Textile and Fiber Engineering (PTFE) merged with the School of Materials Science and Engineering to become what is now The School of Materials Science and Engineering, in Fall 2010. Moreover, two of the authors (MS and JOP) have been funded by the National Science Foundation (NSF) for at least two and a half decades, on various topics dealing with 'soft matter', and it was natural to turn 25 years of research into an educational outcome for undergraduate students.

Most, if not all MSE departments, offer an introductory course on materials. The vast majority of these introductory materials courses are delivered in a strictly lecture format with no laboratory component. Numerous studies confirm that appropriate laboratory experiences significantly enrich and strengthen lecture-based coursework. This is particularly true for laboratory experiences in which students are engaged via hands-on exercises as opposed to demonstrations. Hands-on exercises are challenging to implement for many important materials transformations and analytical methods due to inconvenient timescales, temperature ranges, or equipment access. We have attempted to circumvent these challenges with the use of relatively inexpensive equipment, so that a department/university can afford to offer each student a hands-on experience, and focus on concepts that can be applied across all materials types, that help with the understanding/learning without exposing the students to dangerous or inconvenient temperatures, methods or samples. For many of the experiments described in the text, the equipment will consist primarily of an optical microscope as they are extremely flexible and configurable for a very large variety of experiment types. An additional advantage is that optical microscopes are familiar, being the first scientific instrument to which most of us were introduced. Familiarity should only facilitate the hands-on experience that we envision. The conceptual aspect of the laboratories will be achieved by using soft materials (e.g., liquid crystals, colloidal particles, butterfly wing scales) which will allow many experiments in basic materials principles (e.g., nucleation, crystallization, phase transitions, diffraction) to be conducted at or near room temperature. For example, a laboratory exercise on optical diffraction of colloidal crystals can be used to teach basic concepts of crystal structure as well as *x*-ray diffraction.

When one surveys various textbooks, one finds that there are two types of textbooks. On the one hand, they are works of reference to which students can turn to for the clarification of an obscure point or intimate details of some important experiment coupled with the theory. On the other hand, there are introductory books which deal mainly with principles and help in the understanding of the first type. This book falls in yet another category: one that provides for practical experiments that are relatively simple to carry out on a compound microscope on 'soft materials' with the aim of providing hands-on experience that deals with the principles of materials science. It is not exhaustive by any means, and clearly, it will also become obvious that personal taste plays a key role in the topical coverage of the material. For these sins, we apologize at the very beginning.

The book is divided into two parts, parts I and II. Since most of the experiments are designed to be performed on a compound optical microscope, it was decided that part I will provide as much detail (in non-mathematical terms) as we feel is necessary dealing with the light microscope; from pointing out to why it is necessary to resort to a light microscope, with a historical detour of image formation in a microscope, to a reasonably in-depth discussion of image formation, and resolution of a microscope. This is followed by a discussion of polarization, ways to create polarized light as well as polarized light microscopy coupled with a discussion of conoscopy or optical crystallography. We attempted to provide as complete a background as would be necessary to carry out the experiments that are discussed in part II of the book.

Part II of the book has seven sections or experiments dealing with various aspects of what we believe to be essential for hands-on experiments dealing with the principles of materials science. We start off with a short chapter on how to set up the microscope for observations with and without polarized light. Next there are two experiments dealing with imaging a diffraction grating, both a one-dimensional and a two-dimesional grating. The two-dimensional grating is an interesting experiment where the students make the grating based on what is referred to as 'breath figure templated' assembly from a polymer solution. This has parallels to the classic experiments performed by Bragg and Nye in the late 1940s to illustrate packing and defect formation in metals, and was done using monodisperse soap bubbles, referred to as 'Bubble Rafts'. We also have a short section dealing with colloidal crystal formation and use those to observe diffraction in q-space as well as the structure in real space. This is followed by two sets of experiments dealing with liquid crystals (hardly covered in materials science courses) and the ubiquitous liquid crystal displays. These experiments rely on a good working knowledge of polarized light microscopy, as well as how to use the microscope as a conoscope. This is then followed by an experiment dealing with polymers, as polymers are ubiquitous in our lives. Specifically, the experiment deals with crystallization of polymers at various temperatures and again, the experiments requiring a working knowledge of polarized light microscopy. We end the experiments with optics of butterfly wing scales as an example dealing with optical properties of materials. It is our hope that the book does at least a half decent job of illustrating the utility of a light microscope to interrogate materials referred to as 'soft matter'. Should there be any flaws, I (Mohan Srinivasarao) take full responsibility for them! It should also be noted that this lab course is no longer offered at the GaTech, for a variety of reasons, not the least of which is the short-sightedness of the school of MSE.

Acknowledgments

We would like to thank all the students who underwent the lab course over the span of two years, and the course as well as the written handouts improved thanks in large part to their feedback. Most importantly MS, and JOP acknowledge funding from the National Science Foundation grants over the past two decades. MS would like to thank Professor Peter Collings for the initial reading of some of the experiments, and Professor Oleg Lavrentovich for providing some of the images used in the chapters dealing with liquid crystals. We acknowledge grants a number of grants, DMR-0096240 (CAREER to MS), DMR-0312792, DMR-637233, DMR-0706325, DMR-0907529, DMR-0603026 (MS, JOP) and DMR-2004830 which have contributed to this book.

Author biographies

Mohan Srinivasarao

Dr Mohan Srinivasarao, is a professor in the school of Materials Science and Engineering as well as in the School of Chemistry and Biochemistry, at Georgia Institute of Technology, Atlanta, GA. He has a PhD in chemistry (from Carnegie Mellon University) and works on a variety of problems related to 'soft materials', primarily using light microscopy as a tool. He is a Fellow of the American Physical Society (APS) and of the American Association for the Advancement of Science (AAAS).

Jung Ok Park

Dr Jung Ok Park (PhD in Chemistry from Carnegie Mellon University) is retired from Georgia Institute of Technology, after having worked in School of Materials Science and Engineering (MSE) and the Center for Science and Technology of Advanced Materials and Interfaces (STAMI). Prior to Georgia Tech, she worked at LG Chemical Research Park, in various areas of research and characterization, mostly focused on advanced materials, and holds over 45 patents.

Karthik Nayani

Dr Karthik Nayani, is a professor in the School of Chemical, University of Arkansas, holding the Louis Owen Professorship in Chemical Engineering. He obtained his PhD in Materials Science and Engineering from Georgia Institute of Technology, and is the recipient of the prestigious NSF CAREER award. His lab studies the thermodynamics and dynamic phenomena of liquid–liquid interfaces to design facile biosensors.

Part I

Optical microscopy and optical crystallography

IOP Publishing

Through a Modern Microscope
Optical experiments on soft matter

Mohan Srinivasarao, Jung Ok Park and Karthik Nayani

Chapter 1

Introduction to optical microscope and image formation

1.1 Introduction to light microscopy: the need for a microscope

The world that we inhabit is incredibly beautiful, and this beauty is brought to light with science as a tool. All one has to do is look around to see the beauty that surrounds us. The word '*see*' is the key and forms the basis of this book, where we utilize a light microscope as a tool to explore a variety of materials that can be characterized as soft materials and help to uncover some of the basic principles' behind 'materials science'. One can provide a large number of examples that are beautiful to us, humans, like the pink/reddish color of the clouds in the evening as the sun sets, the incredible northern lights, the breathtaking blue of some oceans, the clear blue of the sky, the brilliant blue seen on the wing scales of a *Morpho* butterfly, the metallic green of the beetle, *Chrysina Glorosia*, a gentle fog over water on a cold morning are among the many phenomena that are fascinating. Figure 1.1 provides an array of such images that we humans call beautiful. All of these are manifestations of light interacting with matter and that interaction detected by the visual system we possess.

Living creatures or organisms, when considered as physical systems, gather information about their environment using a variety of sensory mechanisms such as a sense of smell, sense of touch and a sense of vision, among others. It is, of course, the *sense of vision* that many living organisms critically depend on to gather information about the world, and to act on that information. In human beings, this search for information is carried further due to the insatiable curiosity, and the desire to understand the world that we inhabit. Therefore, when we need detailed information about some parts of the surrounding, it is not surprising that we use our

doi:10.1088/978-0-7503-1431-2ch1

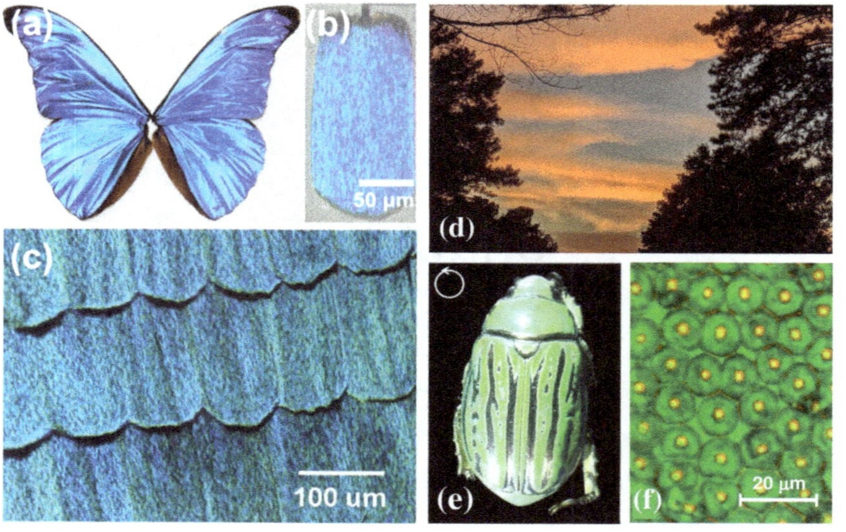

Figure 1.1. Shown are images of what we humans, the visual animals, would call beautiful images. (a) Shows Blue *Morpho* butterfly with its beautiful, metallic blue color, (b) a single scale of the Blue *Morpho*. (c) Shows the tile like arrangement of the individual wing scales of the butterfly Blue *Morpho*. (d) Shows the beautiful, in the late evening near the author's home in Atlanta, USA. (e) Image of a Jewel beetle that reflects circularly polarized light that fascinated Albert Michelson, and (f) shows what the beetle looks like under a polarized light microscope in reflection. All of these images are here to illustrate that the colors are due to the interaction of light with matter as we, humans, are visual creatures. Reproduced from [1] John Wiley & Sons.

eyes to take a very close look at the object. However, the limitations of our visual system do not allow for gathering a far more detailed information about the world that we inhabit, as the objects of interest to us exist in a very wide range of sizes (see figure 1.2).

That the apparent size of any object varies with distance from the observer is well-known, and a common everyday occurrence. To make the size of the object being observed to be larger, we intuitively know that the object must be brought closer to our eyes. Stated differently, the visual angle subtended at the eye by an object depends on its distance. Hence, if an object appears small, then it must subtend a very small visual angle, thus forming an image that is extremely small on the retina; the result is that the image cannot be seen clearly or distinctly. When we bring the small object closer to the eyes, we are in effect increasing the subtended angle, allowing the image to be seen as being distinct or more clearly visible.

The minimum distance between two borders that can be *resolved* is often prescribed as an angle, expressed in minutes of arc. *Visual acuity (*VA) is a measure of how fine the details a person can resolve at maximum *contrast*, and is given by $VA = 1/\alpha$ (arcmin^{-1}), and a value of 1 is considered normal for a 200–300 lx illuminance [2]. So, one can then convert a width measured in centimeters (cm) to an angle measured in degrees by noting that 1 cm at a distance of 57.3 cm from the eye

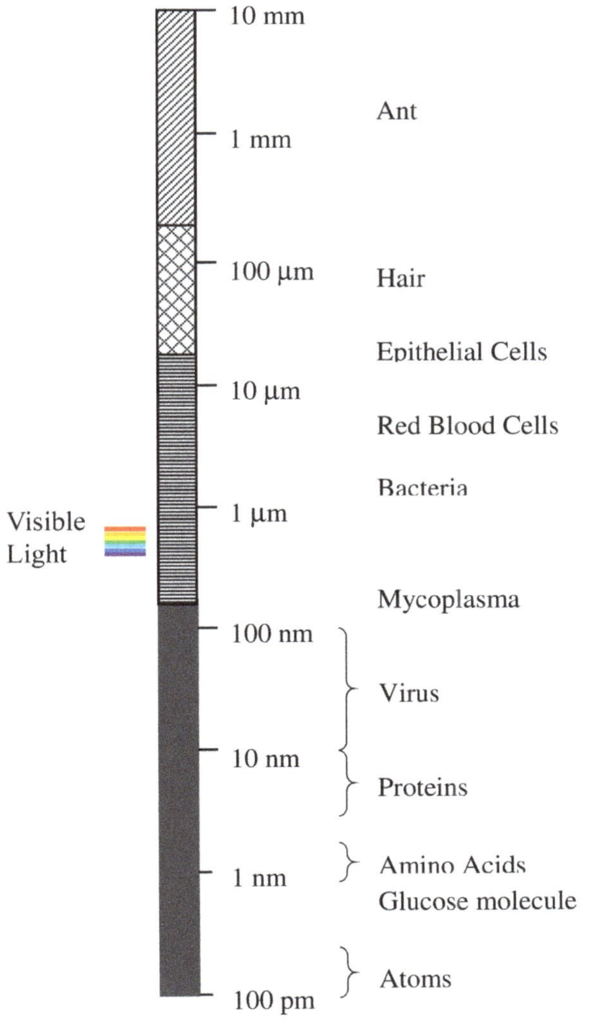

Figure 1.2. The electromagnetic spectrum. The figure shows a logarithmic distance scale (range 1 mm to 0.1 nm). One side shows the wavelength ranges of visible light; for reference, the other side indicates the sizes of various cells and macromolecules. Thus, a red blood cell (7.5 µm) is 15 times larger than a wavelength of visible green light (500 nm). For the eye, the resolution limit (0.1 mm) is taken as the smallest interval in an alternating pattern of black and white bars on a sheet of paper held 25 cm in front of the eye under conditions of bright illumination. Notice that the range of visible wavelengths spans just a small portion of the spectrum. Reproduced from [1] John Wiley & Sons. Copyright 2022 Wiley-VCH GmbH. All rights reserved.

equals 1°. Hence, for an object to be resolved by the human eye, the angle subtended by any object at a given distance ought to be at least one minute of the arc (= 1 arcmin = 1/60°), with some variation. With an acuity value of 1 implies that one can resolve two adjacent lines that are 1 arcmin apart or about 0.17 mm apart at an arm's length from the eye. This is approximately the resolution of the unaided eye.

As the object must be magnified such that in effect the image subtends at least one minute of arc at the eye for it to be resolved, one might suppose that this can be achieved by simply bringing the object closer to the eye, thus increasing its visual angle. However, the power of accommodation of the eye is rather limited, and therefore objects cannot be placed closer than the least distance of distinct vision, if they are to be sharply focused. Therefore, in order for things to be *resolved* when the objects are finer than can be differentiated by the unaided eye, the visual angle ought to be increased in some other way, and it is the *microscope* that allows for the visual angle to be increased.

1.2 Historical tour of light microscopy

It is reasonable to ask, why be concerned with the historical aspects of light microscopy in this book that deals with the utility of microscopy for 'soft materials'. We would argue that it is due to the elegance of the nature of science and the innovations that have fueled the status of light microscopy as an essential tool in many scientific disciplines. It is also because, among all the scientific instruments used by a scientist, an image of someone peering into a microscope most aptly symbolizes the profession (of a scientist) to the nonscientist. Sir Hans Krebs, Nobel Laureate in Physiology or Medicine (1953) [3] provides an answer: '*Those ignorant of the historical development of science are not likely ever to understand fully the nature of science and scientific research*'. Since the history of light microscopy or optical microscopy or imaging in general, is particularly rich in beautiful experiments (and elegant theories), it would appear worthwhile to take a short detour into the pages of history dealing with the development of microscopy as the essential tool that it has become.

1.2.1 Historical aspects of microscopy

Historically, the need to 'see' things better has led to the development of tools that have accomplished that purpose, the optical microscope and the telescope being two such examples. The earliest of imaging devices was a glass lens found in the ruins of Nineveh, dating back to 610 BC. A Roman philosopher later (about 100 AD) noted that a globe of water held properly would magnify objects, when Seneca the Younger (3 BC–65 AD) wrote: 'Letters, however small and indistinct, are seen enlarged and more clearly through a globe or glass filled with water.' It appears that the earliest examples of a single-lens used as reading magnifiers date back to the 12th century. The first simple microscope appears to have originated in the first part of the 16th century consisting of a short tube with a lens at one end, and a flat glass at the other end to place the object to be examined. Another version of a simple microscope consisting of a hand-held lens with a spike in front to hold the specimen was invented. This invention, a simple microscope, a natural world unseen by the naked eye was brought to light. In the early part of the 17th and 18th centuries, the simple or single-lens microscope was a far more useful instrument for scientific observations than the compound microscope (consisting of two lenses) as the compound microscopes suffered from chromatic and spherical aberrations. These problems were eliminated when improved lenses (made of two glasses) were developed that made compound microscopes scientifically useful.

Antonie Philips van Leeuwenhoek (1632–1723), a Dutch tradesman and scientist, was best known for his work on the improvement of the microscope as well as his contributions towards the establishment of microbiology. In 1674, he started to observe bacteria and protozoa, his 'very little animalcules', isolated from various sources, such as rainwater, pond water and well water among other sources such as the human mouth. Van Leeuwenhoek's studies included the microscopic structures of seeds, bones, skin, fish scales, insect eyes, muscle fibers and 'animalcula'—the microorganisms. Much of the work done by van Leeuwenhoek was done with simple microscopes that are extremely simple devices, consisting of a single, high-quality lens of very short focal length. His microscopes could magnify up to 275× and resolve down to 2 μm. Brian Ford, in a delightful article entitled 'Did physics matter to the pioneers of microscopy ?'[4], notes that 'Simple microscopes are small and inexpensive and can produce images that compare surprisingly well with those obtained with present-day instruments'. He also notes that 'Simple microscopes from several centuries ago can perform better than a modern instrument in unskilled hands'. Historically, it should be noted that many have minimized the work performed by the early pioneers in microscopy, many of whom used a simple microscope. As an example of that dismissal, Ford [4] cites a talk at the American Physical Society by Daniel Deutsch entitled '*Did Robert Brown observe Brownian Motion: probably not*' and the title clearly negates the need to read the article. Motivated by this controversy of sorts, Ford, in a later work did in fact observe the constant agitation of particles, that is now referred to as 'Brownian Motion' using a simple microscope [5]. Ford also notes, that the early pioneers of microscopy worked by trial and error, and were not constrained by theoretical physics nor burdened by the perceived limits of resolution of the microscopes they used. Figure 1.3 illustrates the evolution of the microscope over the past four centuries, which displays a simple microscope of the type used by van Leeuwenhoek, a compound microscope fashioned by Robert Hooke, a polarizing microscope (~1880) and a current, state-of-the art Leica microscope.

The optical microscope itself dates back to the early 1600s, and was primarily considered a *philosophical* instrument. It is believed that the optical microscope made its transformation from a philosophical instrument to a research/diagnostic tool around the 1840s. This was primarily due to the increased effort to improve public health as a result of the outbreak of cholera epidemics in Britain around the 1830s [7]. The optical microscope still continues to play a very important role as a research/diagnostic tool and the picture of a 'researcher' peering into a microscope has become a symbolic representation of the sciences, in general.

The 1830s and 1840s also saw an increase in the quality of the microscope in terms of small features that could be seen [7]. The fact that smaller and smaller features were resolved with progression of time is shown graphically in figure 1.4. This graph clearly illustrates the increase in resolution, reaching its limit (see the next section) around 1880 [7]. This coincided with the large increase in the manufacture of microscopes during the same period. The limit of resolution

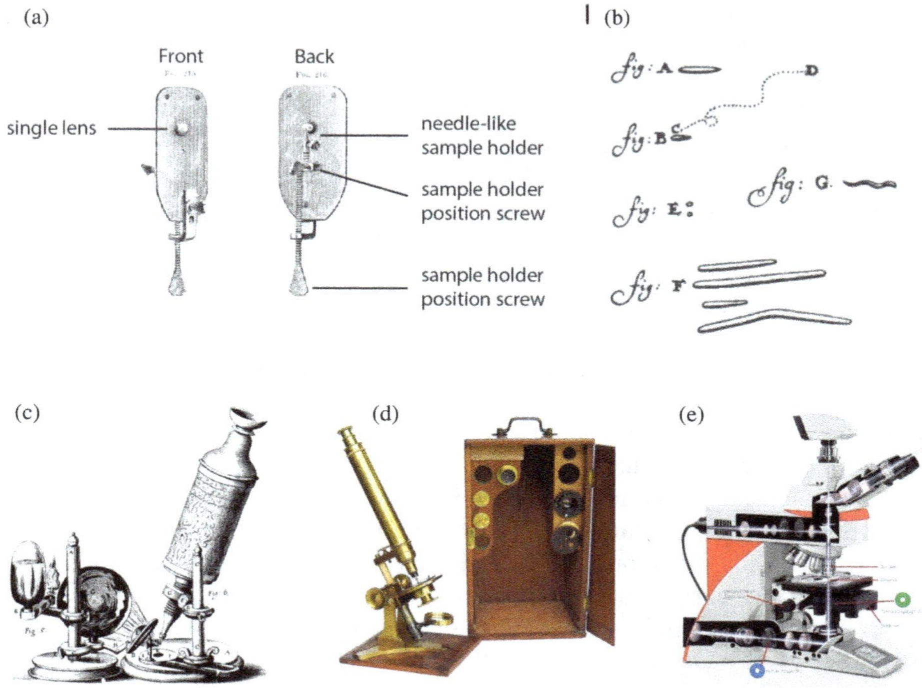

Figure 1.3. van Leeuwenhoek's microscope and its observations. (a) Hand-made single-lens microscope with mechanically moving sample holder. Adapted from [6]. (b) Observations of human mouth bacteria, as described by van Leeuwenhoek in his 39th letter to the Royal Society (1683). Visible are (A) *Bacillus*, (B) *S. sputigena*, (C) and (D) the path it makes, (E) *Micrococci*, (F) *L. buccalis*, and (G) a spirochete. This Leuwenhoek picture of animacules image has been obtained by the authors from the Wikimedia website where it was made available by Huydang2910 under a CC BY-SA 4.0 licence. It is included within this book on that basis. It is attributed to Huydang2910. (c) Robert Hooke's microscope. This Hooke-microscope has been obtained by the authors from the Wikimedia website, where it is stated to have been released into the public domain. It is included within this book on that basis. (d) John Browning polarizing microscope (~1880). Image credit: Peabody Museum of Natural History. (e) state of the art polarizing light microscope (courtesy Leica Microsystems).

(the ability to distinguish closely spaced objects) of an optical microscope was recognized by Abbé with his fundamental work on diffraction and image formation [8]. He recognized that light of wavelength λ cannot be focused to a size below $\lambda/2$, which in turn implies that one cannot distinguish two particles that are spaced less than $\lambda/2$ apart. This implies that if one uses green light, then the resolution is about 0.25 μm. This resolution was obtained as far back as 1880 with the use of oil immersion objectives, as is clearly seen from figure 1.5. We will come back to the issue of resolution in later sections, since this is important to understand in the context of microscopy. However, we are not going to be too concerned with a revolution in increasing the resolution that has taken place over the past 30 to 40 years, resulting in the 2014 Nobel Prize in Chemistry for '*for the*

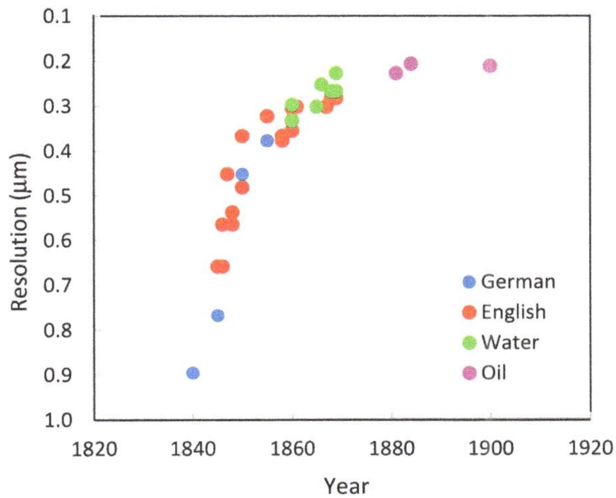

Figure 1.4. Illustration of the improvements in the resolution of microscope objectives as a function of time. Data from G L'E Turner, [7]. Reproduced from [1] John Wiley & Sons.

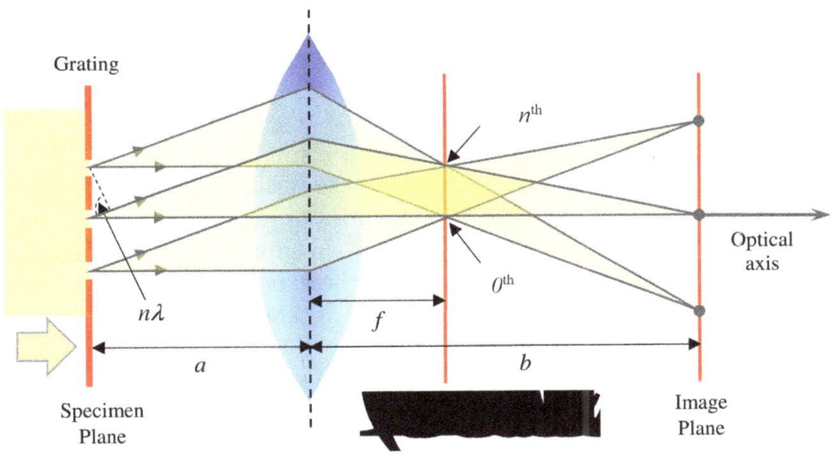

Figure 1.5. A schematic diagram showing image formation in a microscope, showing Abbé's theory for image formation in an optical microscope. Reproduced from [1] John Wiley & Sons.

development of super-resolved fluorescence microscopy' [9]. The term *super-resolution* is often used be to describe any (optical) imaging technique that is able to produce resolution beyond this diffraction limit of $\lambda/2$.

1.3 Image formation in a microscope and resolution

The term 'light microscopy' in its simplest form involves passing visible light that is transmitted through or reflected from a sample through a single lens (simple microscope) or multiple lenses (a compound microscope) to allow for a magnified view of a given sample. The compound optical microscope is an instrument that uses visible light to produce a magnified image of the object that is projected onto the retina of the eyes or an imaging device. As is perhaps obvious, the word 'compound' refers to the use of two lenses, the microscope objective and the eyepiece (the ocular) to provide for a magnified image of the object.

A detailed description on image formation in a microscope can be found in many excellent references [8, 10], hence in this section we provide a brief account of how an image is formed and what is the limit on resolution. It should be recognized that there is a clear distinction between resolving an image and visualizing an object. Visualization of objects that are far below the resolution limit of an optical microscope, the so-called 'diffraction limit', is not a new event [10] but resolution below the diffraction limit is [9, 11]. These two issues must not be confused with one another.

The issue of resolution and what is attainable by an imaging system was tackled by Ernst Abbé and the limit to the resolution has been termed the Abbé limit or the diffraction limit. Abbé's first publication in 1873 in German (a translation can be found in a book by Barry Masters [12]), is merely a verbal treatment of the diffraction theory of microscopic image formation, the sine condition, but without mathematical derivations. Pertaining to the diffraction theory of microscopic imaging, Ernst Abbé states: 'The limit of discrimination will never appreciably exceed a whole wavelength under central illumination ... and half a wavelength ... under extreme oblique illumination' [13]. This statement thus implies, $\Delta = \lambda/(n \sin \sigma)$ or $\Delta \geqslant \lambda/(2n \sin \sigma)$. Quite remarkable!

At this point it might be worthwhile, from a historical perspective, to note that Abbé was reluctant to publish his discoveries in the scientific literature but published in journals that catered to the use and applications of microscopes as his motivation was the promotion of new Zeiss microscopes. It appears that his disposition towards publishing changed when Abbé learned that he been nominated for the Nobel Prize in Physics 1904 [12], leading to the publication of Abbé's collected works, the first volume dealing with the theory of microscope.

Albert B Porter in 1906, writing 'On the diffraction theory of microscopic vision' [14] writes about Abbé's theory rather succinctly in the introduction, and states:

> *'Although thirty-two years have passed since Professor Ernst Abbé proposed his diffraction theory of microscopic vision, it is still to some extent a matter of controversy among microscopists, and is perhaps less familiar to physicists than its importance warrants'.* Then Porte summarizes Abbé's theory thus: *'If a lens is to produce a truthful image of an illuminated object, it must have an aperture sufficient to transmit the whole of the diffraction pattern produced by the object; if but parts of this diffraction pattern is transmitted, the image will not truthfully represent the object, but will correspond to another (virtual) object whose diffraction pattern is identical with that portion which passes through the lens; of the structure of the object is so fine, or if the aperture of the lens is so narrow, that no part of the diffraction pattern due to the structure is transmitted by*

> *the lens, then the structure will be invisible no matter what magnification is used. Abbé and others have devised a number of interesting experiments to illustrate the theory, but the complete mathematical development has never been published'.* Then Porter proceeds with a mathematical theory stating *'The particular case in which the object is a transmission grating consisting of alternate opaque and transparent lines may, however, be treated by means of a simple application of Fourier's theorem'.*

Thus began the full-fledged development of mathematical formulation of Abbé's brilliant experiments.

The Rayleigh criterion is the distance between two just-resolvable points. It is often used to describe the resolution of optical instruments and image-forming devices. There have been several discussions on the resolution limit, claiming that the Rayleigh limit is only a convenient measure of the resolution obtainable by optical instruments, and that it should be possible, in principle, to obtain resolution without limit [15]. In an elegant paper, G Toraldo di Francia [16] showed that even though it should be possible to obtain resolution without any limits based on mathematical grounds, the practical realization of the mathematical ideal is quite impossible. Hence, it is worthwhile to remind ourselves of the arguments that have been presented.

We begin by considering an imaging system schematically shown in figure 1.5, as envisioned by Abbé. Image formation in a microscope may be understood in terms of diffraction of the incident beam on passing through the sample and recombination thereafter to give a real (and magnified) image of the object [8].

Image formation in a microscope can then be considered as a double diffraction process. In order to illustrate this in a qualitative fashion, let's consider an object which is a diffraction grating. A parallel beam of light, on passing through this object, produces several orders of diffraction in the back focal plane (BFP) of the objective (first lens). The diffracted beams emerge at angles θ_ι, given by $nd \sin \theta_\iota = m\lambda$, where n is the refractive index, d is the grating period of the object, m is an integer, and λ is the wavelength of light. Each diffracted order can be considered as a plane wave and the set of plane waves produced by the diffraction grating can be collected by a lens (or an objective) having a sufficiently large numerical aperture (NA), so that they converge individually to a set of points in the focal plane (Fourier transform plane) of the objective. The diffracted light passes on to interfere, resulting in an *accurate* image of the diffraction grating, only in a plane conjugate to that of the grating (i.e., the object), which is the image plane.

A diffraction grating was chosen as the object for several reasons. Historically, a grating was used to demonstrate Abbé's theory [14, 15] but it also serves the purpose of illustrating that image formation in a microscope can be considered as two distinct stages: (a) production of the Fraunhofer diffraction pattern in the rear focal plane and (b) a Fourier transform of the Fraunhofer diffraction pattern to form a magnified image. In the focal plane of the lens, having produced the Fraunhofer diffraction pattern of the object, the various diffraction orders behave like a set of equally spaced point sources and the image is their diffraction pattern. The finest structure observable is determined by the highest order of diffraction that is transmitted by the lens used for the purpose of imaging, and is a function of the numerical aperture of the objective used in the microscope. The *numerical aperture* (NA) of an objective is the light gathering power and is defined as $NA = n \sin \theta$, where n is the refractive index of the medium in which the light is collected and θ is the half-angle of the cone of light entering the objective. It is a dimensionless number that characterizes the range of angles that the imaging system (in this case a microscope objective) can accept, as is evident.

In a light microscope, the part of the Fraunhofer diffraction most likely to be lost is the outermost part, which corresponds to larger angles of diffraction, and it will be lost if the numerical aperture (the light gathering power) of the objective is too small. This results in a loss of sharpness in the image, a deterioration of the resolution and a loss of the finer detail, since it is the sharper boundaries and finer detail that diffract light at larger angles. It is seen that it is useful to collect as many orders of diffraction as possible, since this contributes to the detailed structure of the image.

In order to better understand the image formation, one can resort again to Fourier optics [15]. Consider the image of a grating as described above. In the back focal plane (which is the Fourier plane), we have several orders of diffraction behaving as a set of equally spaced point sources. The zero-order beam in the diffraction plane produces a uniform amplitude or uniform illumination across the image plane, which by itself carries no information about the period of the grating. Each higher order beam produces a sinusoidal amplitude distribution. These various amplitudes are superposed in the image plane to produce the image. The amplitude distribution across the image plane can then be represented in terms of various Fourier components or *spatial frequencies.* To observe the periodicity of the grating, it is enough to reproduce the fundamental (or the lowest spatial frequency) component in the amplitude distribution of the image. This is accomplished if the objective of the microscope can admit the zero order and *one* of the first-order beams, giving two effective sources in the Fourier plane, thus giving maxima in the intensity distribution across the image plane which correspond to the grating spaces, resulting in a resolved grating. The process of formation of an accurate representation of the object is schematically illustrated in figure 1.6, where the conditions for the formation of an image as well as for not forming an accurate representation of the object are shown. In figures 1.6(a) and (b), it is apparent that only the zero-order beam is captured by the objective and hence, no image is formed of the object. In figure 1.6(c) the zeroth-order as well as the first-order beam are captured, thus leading to the formation of an image, while in figure 1.6(d), multiple orders are captured by the objective, thus leading to a greater definition of the image of the object.

We are now in a position to state the *Abbé principle: this principle states that both the zero-order and one of the first-order beams must be transmitted by the objective.* The limit to resolution is then given by the finest grating that can be resolved in this way. It becomes clear that finer gratings give rise to larger angles between the zero and first orders, thus requiring lenses of larger numerical aperture in order to be resolved. Hence the resolution of a microscope is a function of the numerical aperture of the objective, and the theoretical limit is given by the expression of $0.61\lambda/NA$ [13, 15]. This expression is often approximated by $0.5\lambda/NA$ and comes about

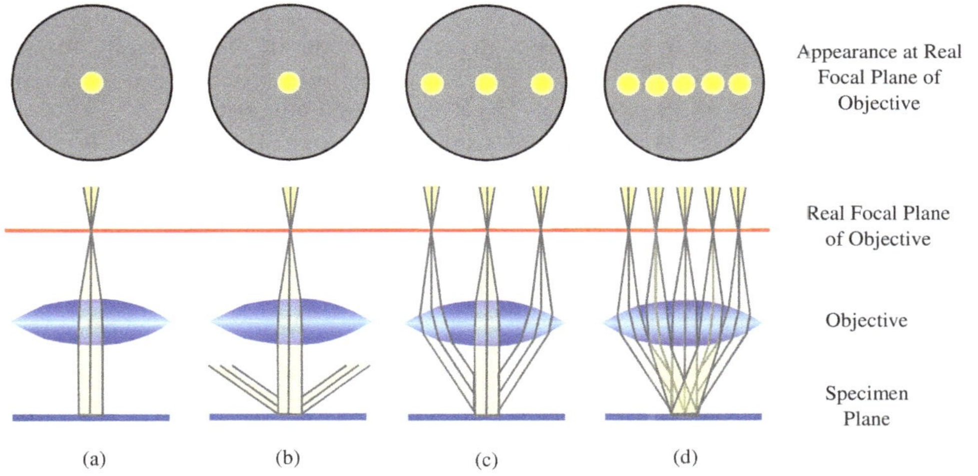

Figure 1.6. A schematic illustrating the generation of an image by interference. In order for interference to occur one needs to collect two adjacent orders of diffracted light. In (a), no diffraction occurs and in (b), the diffracted rays are not collected by the objective. In (c), a minimum of two orders (0th and 1st) are collected leading to an image. In (d), multiple orders are captured, leading to a high degree of definition in the image. Reproduced from [1] John Wiley & Sons. Copyright 2022 Wiley-VCH GmbH. All rights reserved.

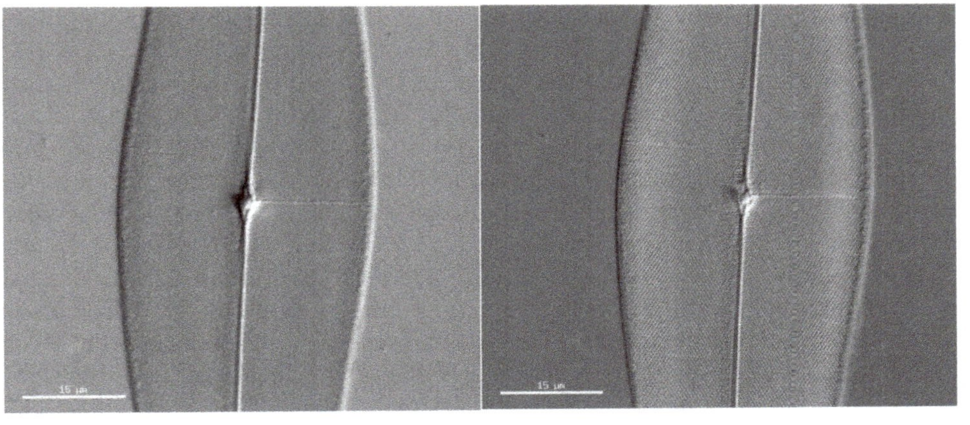

Figure 1.7. Images of diatoms with an Olympus 40× UAPO objective with an adjustable aperture varying from 0.65 to 1.35. Image on the left (unresolved) is at an *NA* of 0.65 and the nicely resolved one on the right is with an *NA* of 1.35. It is very clear that higher *NA* leads to much better resolution, something that Ernst Abbé demonstrated in 1873 with a similar diatom. Source: Courtesy of Dr Alison North of the Johns Hopkins University.

when one takes account of the numerical aperture of the condenser. Further details can be found elsewhere [10].

In an effort to demonstrate that the numerical aperture plays a central role in imaging at high resolution, images of a diatom at two different numerical apertures are shown in figure 1.7. The image on the left is with an *NA* of 0.65 and the one on the right is with an *NA* of 1.35. It is evident that the one on the right is better resolved. These two images were very generously generated, specifically for this chapter, by Dr Alison North, a Senior Director of the Bio-Imaging Resource Center and Research Associate Professor, at The Rockefeller University.

The above discussion represents an approximate theory of image formation. It is assumed that light associated with a given diffraction order is in a single direction and all light associated with this order is either accepted or rejected by the collecting objective. The theory does not take into account of the possibility that part of the light in a given diffracted order is admitted and the finite amount of light is diffracted between the principal maxima.

The above discussion was developed for nonluminous objects, as Abbé described his theory for the microscope. Rayleigh [17] took a different approach to the problem. Light from each point in an object that is illuminated by a condenser produces a set of circular rings (called the Airy pattern, see figure 1.8) in the conjugate image plane, and the final image is a superposition of these Airy patterns. The effects of such superposition can be calculated if the phase relations and the degree of coherence are known, which are determined by the structure of the object and the form of illumination. The Airy pattern results from diffraction by the objective and the structure of the object is taken into account by the phase relations between the Airy patterns. This leads again to the definition of resolution, the *Rayleigh criterion*, the separation of two just-resolvable points. This is given when the maxima of one Airy pattern is located at the minimum of the other Airy pattern. The resultant two-point intensity will then have a minimum, between two just-resolved objects, that has about 20% less intensity than the maximum. In all of this, nothing has been said about signal-to-noise (S/N) ratio in the measurements. As with any measurement, S/N ratio plays an important role. Hence, if one can have a good S/N ratio in the measurement of the convoluted intensity distribution from two just-resolved Airy patterns, a smaller separation can be estimated with adequate certainty.

As discussed above, Lord Rayleigh noted that two-point light sources of identical intensity could be resolved by the human eye if the Airy disk of one point does not come closer than the first minimum of the second point's Airy function (figure 1.9). When this condition is met, a visible decrease of intensity of around 20% will occur between two partially overlapping Airy disks. It should be noted that '*clearly be resolved by the human eye*' is of course a very subjective measure, and therefore Hecht and Zajac noted in their popular textbook book *Optics* [19]:

> '*We can certainly do a bit better than this, but Rayleigh's criterion, however arbitrary, has the virtue of being particularly uncomplicated*'.

We now return to the issue of resolution and its limits. It is known that an object of finite extent has a Fourier transform that is an analytic function. Given enough information, even if a part of the Fourier transform is lost (due to the presence of

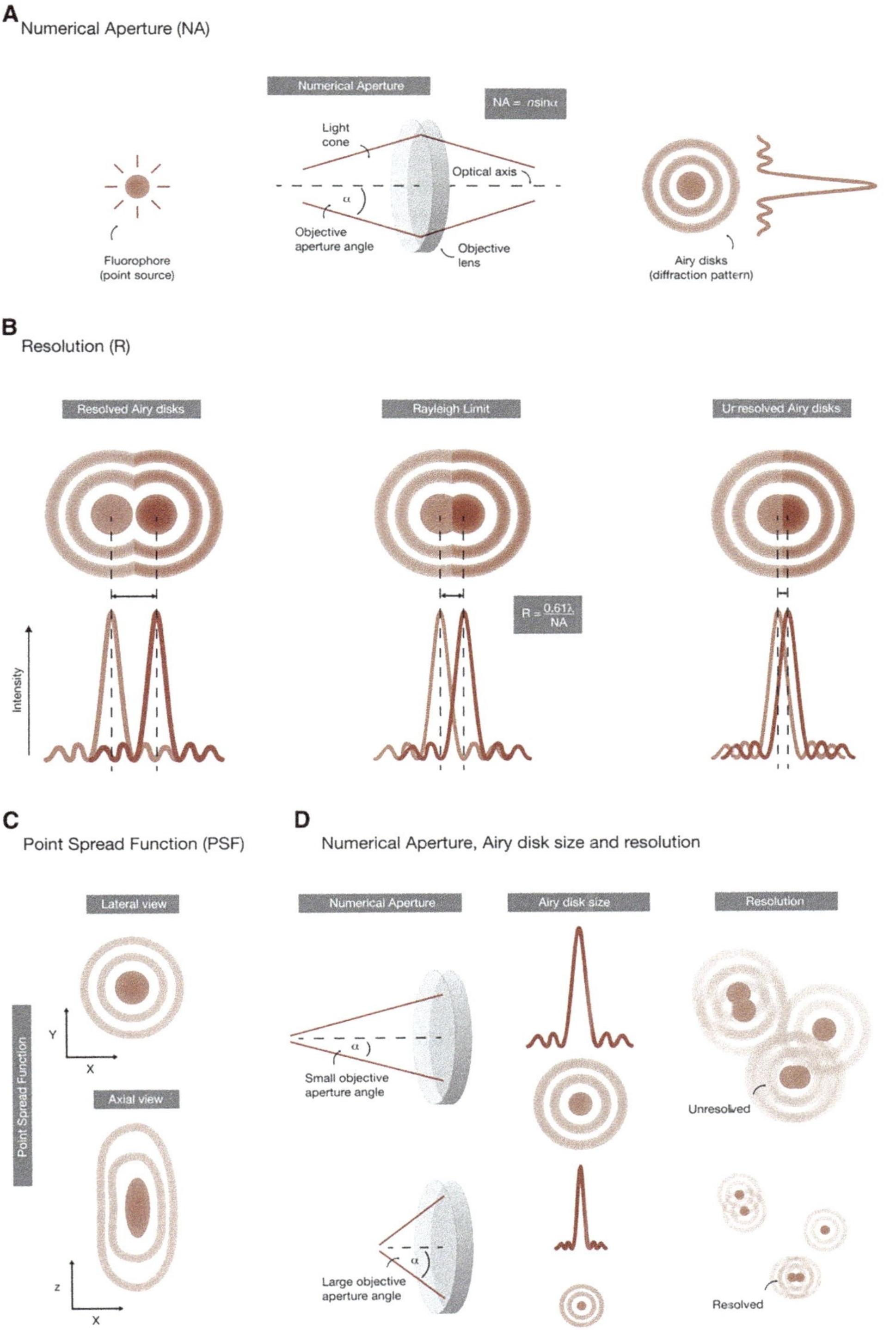

Figure 1.8. (A) Fluorophores can be considered as point sources of light that emit photons. Due to the wave nature of light, emitted photons are diffracted at the optical components of the microscope and eventually appear as regularly spaced rings (Airy disks) at the detector unit. The number of photons that

can be collected by the objective depends on the size of the objective aperture angle (α) and the refractive index of the sample immersion medium (n). Together, both values define the NA ($NA = n \sin \alpha$) of an objective, which is a dimensionless measure that describes its performance. The NA of objectives that are used for fluorescence microscopy typically ranges from 0.2 to 0.95 for air objectives and 0.85–1.4 for oil objectives. (B) The ability of a fluorescence microscope to resolve two-point sources of light that are in close proximity is essentially defined by the Rayleigh criterion ($R = 0.61\lambda/NA$). It states that the maximum resolution of a fluorescence image is roughly half the emission wavelength of the fluorophore. When two fluorophores are further apart they can still be resolved, while fluorophores that are too close to each other appear as a single spot. (C) The point spread function (PSF) is the three-dimensional representation of the Airy pattern and a unique fingerprint of each imaging system. In the lateral dimension, the PSF appears as regular rings, while in the axial dimension these rings appear as elongated ellipsoids. Hence, the shape of the PSF resembles that of an hourglass. (D) The NA of the optical system fundamentally defines its capability to resolve details in the image. Larger collection angles generate smaller Airy disks of individual fluorophores at the detector and consequently produce images with greater resolution. Reproduced with permission from [18] Creative Commons Attribution 4.0.

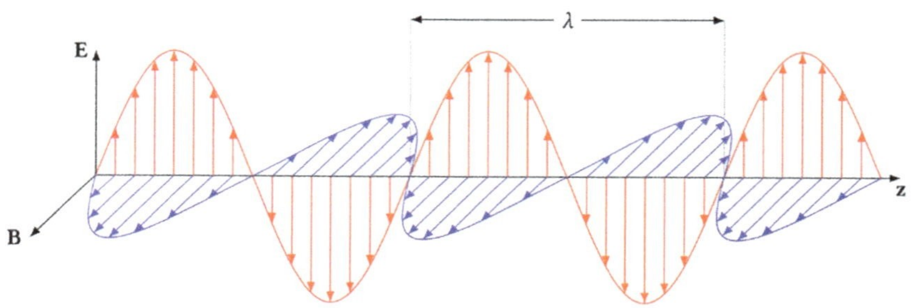

Figure 1.9. A 'vertically polarized' electromagnetic wave of wavelength λ has its electric field vector **E** (red) oscillating in the vertical direction. The magnetic field **B** (or **H**) is always at right angles to it (blue), and both are perpendicular to the direction of propagation (z). This Electromagnetic wave2 image https://en.wikipedia.org/wiki/Polarization_(waves)# has been obtained by the authors from the Wikimedia website where it was made available by Francois~frwiki under a CC BY-SA 4.0 licence. It is included within this book on that basis. It is attributed to Francois~frwiki.

effective apertures) in the Fourier transform plane (back focal plane), it must, in principle, be possible to recover the entire function. This has been the topic of many elegant papers [15, 16], presenting arguments that the Rayleigh criterion is just a practical limit and not a fundamental one. This argument is mathematically quite correct, as was pointed out by Goodman [15], among others. It should be recognized that these arguments are of little use in terms of going around the diffraction limit, due to the issue of noise entering into the detection scheme. It is inevitable that one will be limited by the noise in any measurement or detection scheme, and hence there is really very little hope of being able to precisely reproduce the object field from a knowledge of the image field. Nonetheless, the Rayleigh criterion is a useful limit of resolution to think about when considering any scheme for enhancement of lateral resolution.

Thus far the discussion has been on lateral resolution. Since optical sectioning microscopy (OSM) has revolutionized biology, there has been considerable effort to obtain thinner and thinner sections, and a number of methods have been developed in order to achieve high axial resolution. This form of microscopy has been primarily used in biology to study the structure of cells in three-dimensions. The image formed in an optical microscope contains information from the area that is in clear focus in a thin focal plane, as well as out-of-focus information. This is especially true if the object is much thicker than the depth-of-field. The depth-of-field can be as low as 0.15–0.20 μm when a high magnification high NA objective is used. A vast majority of experiments use fluorescence microscopy to visualize the 3D structure of cells, and the imaging is usually done in low light conditions. This decreases the S/N ratio and limits the axial resolution attainable.

Axial resolution is defined as the least distance between two objects separated only axially (along the optic axis of the microscope), which can be identified *not to be* in the same plane (plane normal to the optic axis of the microscope). As there is no single agreed upon formula that defines the limit to axial resolution, we use the expression given by Francon [20] for the axial setting:

$$\Delta z = \frac{\lambda}{4n \sin^2(\theta/2)}$$

where Δz is defined as the minimum axial separation when two objects can be distinguished as such by changing the focal plane. This expression would predict an axial setting accuracy or resolution of about 0.18 μm to 0.14 μm for high NA lenses ($NA = 1.4$), which has been demonstrated by Inoue [21].

1.3.1 Abbé's diffraction experiments parts 1–5, by Peter Evennett, Dresden Imaging, Facility Network, 2001

Dr Jeremy Sanderson, Bio-imaging Facility Manager at MRC Harwell Institute in the UK, alerted me to the YouTube videos produced by Peter Evennett of the Dresden Imaging Facility Network 2001, demonstrating Abbé's experiments to provide an understanding of image formation in a light microscope. These are terrific videos and done immaculately, and I highly recommend anyone to view these five videos to understand image formation in a microscope, and how to use the instrument properly to do imaging of the objects that one wishes to image. These are narrated in English, and they are remarkable for the clarity with which Abbé's theory of image formation is described, and anyone wanting to understand microscopy as an underutilized and widely abused instrument, would do well to pay attention to these videos.

- Abbe's Diffraction Experiments Pt. 1 (9:35 min) by Peter Evennett, Dresden Imaging Facility Network: https://www.youtube.com/watch?v=fAuo7NIS97U (accessed March 30, 2021).

- Abbe's Diffraction Experiments Pt. 2 (9:38 min) by Peter Evennett, Dresden Imaging Facility Network: https://www.youtube.com/watch?v=rPAJ2Vs9A_I (accessed March 30, 2021).
- Abbe's Diffraction Experiments Pt. 3 (9:38 min) by Peter Evennett, Dresden Imaging Facility Network: https://www.youtube.com/watch?v=CbPoi9bie1I (accessed March 30, 2021).
- Abbe's Diffraction Experiments Pt. 4 (9:43 min) by Peter Evennett, Dresden Imaging Facility Network: https://www.youtube.com/watch?v=wZ369WKiJdQ (accessed March 30, 2021).
- Abbe's Diffraction Experiments Pt. 5 (3:48 min) by Peter Evennett, Dresden Imaging Facility Network: https://www.youtube.com/watch?v=t6RO7fusHYk (accessed March 30, 2021).

1.3.2 How to use a microscope

It has been our experience, and noted by many other authors, that most users of microscopes learn the utility of the instrument by trial and error long before reading a book or a chapter on the subject of microscopy. For example, in the preface to the book on *Optical Microscopy of Materials*, R Haynes states: 'Despite its extensive use, it has been described as the most misused, abused, and misunderstood scientific instrument, for it will produce an image of a sort no matter how badly it is used. To use it effectively, even in its simplest applications, a knowledge of the simple theory of the microscope is necessary, for theory shows and explain how it should be used'. Jeremey Sanderson in his book on *Understanding Light Microscopy* writes that 'practicing microscopy appears simple, but scientific journals are replete with examples of poor-quality images. Imaging is more than just "the bit tacked on the end of the experiment"'. He quotes Helen Pearson [22], who notes:

> *To correctly capture images using a modern microscope, researchers must have a good grasp of optics, an awareness of the microscope's complexity … such skills can take months or even years to master [which] owing to inexperience or the rush to publish, are all too often squeezed into hours or days. … Inept microscopy, and subsequent analysis, can easily generate results that are misleading or wrong.*

Last but not least, Henry Baker in his book *The Microscope made Easy and Pocket Microscopes* (1742) had the following advice for user of microscopy in chapter 15, entitled 'Cautions in Viewing Objects', and still applicable today, even though the microscopes are so much more advanced than the time when this warning was issued. The warning from Baker:

CHAP. XV.

Cautions in Viewing OBJECTS.

BEware of determining and declaring your Opinion ſuddenly on any Object; for Imagination often gets the Start of Judgment, and makes People believe they ſee Things, which better Obſervations will convince them could not poſſibly be ſeen: therefore aſſert nothing till after repeated Experiments and Examinations, in all Lights, and in all Poſitions.

When you employ the Microſcope, ſhake off all Prejudice, nor harbour any favourite Opinions; for, if you do, 'tis not unlikely Fancy will betray you into Error, and make you think you ſee what you would wiſh to ſee.

Remember that Truth alone is the Matter you are in ſearch after; and if you have been miſtaken, let not Vanity ſeduce you to perſiſt in your Miſtake.

For the microscope to be used effectively, whether it is for visual observation or particularly for photomicrography, it is self-evident that the microscopic object being interrogated be illuminated uniformly across the field of view. As Peter Evennett notes in his paper on Köhler Illumination, the simplest way of having an evenly illuminated field of view was to 'throw on to the object an image of another area which is uniformly illuminated'. He notes, in Köhler's time this was a large white cloud in the sky, or by night the flame of an oil-lamp, imaged onto to the specimen by means of a lens. This lens now is called the 'condenser'. Instead of this so-called 'source-focused' illumination, we now use the innovation due to Köhler. Köhler introduced an additional lens, now known as the lamp collector lens, and moved the light source to the back, so that an image of the source was projected into the first focal plane of the condenser. This, Evennett notes, 'is essentially similar to having the light source itself in this position, and makes it possible to include a diaphragm for regulating the illuminating aperture at here'.

Before using a microscope for imaging, one always ought to set it up for Köhler illumination, as it provides for a uniformly illuminated field of view. The following is a step-by-step instruction of sorts, to set the microscope for Köhler illumination. This optical arrangement assumes a matched numerical aperture for the objective and the sub-stage condenser. Mismatch with a larger numerical aperture for the objective results in flare in the image, while the reverse results in decreasing illumination toward the periphery of the image, and degraded optical resolution. To adjust the microscope for Köhler illumination, the following procedure ought to be adequate:

- Center and focus the light source image projected into the condenser. In other words, the tungsten light source should be aligned along the optic axis of the microscope to provide for an even illumination at the stage.
- A microscope slide with a high-contrast object is placed on the stage and brought to focus with a 10× objective. The condenser diaphragm is closed down and the height of the condenser is adjusted until a sharp image of the condenser diaphragm is superimposed on the object on the microscope stage. This image is then centered in the field of view using the centering screws provided in the sub-stage condenser.
- To limit the flare, the field diaphragm (immediately below the converging lens in the sub-stage condenser) is adjusted so that its edges are visible in the rear focal plane. Placing the Bertrand lens in the optical path brings the rear focal plane of the objective into the field of view for this adjustment.
- Align the laser. The laser is aligned to be in the center of the field of view. (Do not view image unless between crossed polars or otherwise attenuated.)
- Repeat steps with 32× objective ($NA = 0.60$), as well as with other objectives when they need to be used.

Following this procedure ensures that the optical elements will be aligned along the optical axis of the microscope, providing a homogeneously illuminated field of view. In order to use the microscope as a conoscope, a He–Ne laser with a wavelength of 633 nm can be added to the microscope as a light source. This is necessary, since samples possessing large retardation will not give rise to an interference figure in the rear focal plane, if only an interference filter is used to select a wavelength of white light owing to the remaining wavelength dispersion in the incident beam. Also, when using a laser source, the laser speckles need to be eliminated by having a diffuser in the path of the beam.

1.4 Polarized light microscopy

1.4.1 Polarized light

Polarized light microscopy is a common tool used to study a variety of materials that are considered to be anisotropic or ordered materials, with liquid crystals, polymeric materials that are semicrystalline or polymeric fibers as examples. The technique exploits the optically anisotropic properties of the material to create a contrast—the contrast then depends on the interaction of polarized light with the specimen or sample of choice and the outgoing light is then analyzed using another polarizer, often referred to as the analyzer, as it is useful for the analysis of the outgoing light beam.

A polarized light microscope is basically a compound microscope with the addition of two polarizers in the path of light beam. One polarizer is located below the sample stage, and is often simply referred to as the polarizer. The second, located above the stage, is often simply referred to as the analyzer, as it 'analyses' the light that exits from the sample located on the stage. In most modern microscopes the polarizer transmits plane-polarized light or linearly polarized light vibrating in the North–South direction, and the analyzer allows only light (or a component thereof) which vibrates in the East–West direction.

Should an object be viewed with the analyzer inserted in the path, then the sample is viewed under crossed polarizers or crossed polars. If the analyzer's privileged direction is rotated by 90° to be parallel with that of the polarizer, then the sample is said to be viewed under parallel polarizers or parallel polars. If the analyzer is not inserted at all, then the sample or object is said to be observed with plane-polarized light, with the plane of polarization being provided by the illuminating light passing through the polarizer below the stage.

By now it must be obvious that light is a transverse electromagnetic wave, implying that light consists of a coupled oscillating electric field and magnetic field that are always perpendicular to each other as well as being perpendicular to the direction of propagation of the light beam. Hence, when dealing with what is referred to as a plane-polarized light beam, one is referring to is the relationships between the directions of the fields and the direction of propagation. For the most part when dealing with interaction of light with matter, the oscillating magnetic field is ignored because the electric field has a greater propensity to interact with matter. When a beam is said to be linearly plane-polarized, the electric field (denoted by a vector) oscillates in a single direction perpendicular to the direction of propagation. Often, this is also referred to as the plane of polarization and is the plane that contains the electric field vector and the direction of propagation. In the case of circular or elliptical polarization, the fields rotate around the direction of propagation at a constant rate; and of course, there being no bias, the rotation can be either in the left-hand or right-hand direction. However, we will primarily deal with linear or plane-polarized light beams for the discussion dealing with anisotropic materials.

Most of the normal sources of light, like the sun, the flames and incandescent lamps produce light that is referred to as *unpolarized light* or *randomly polarized light*. This implies that the electric field vector that describes the light beam, vibrates in all directions, albeit perpendicular to the propagation direction. There are a number of ways to produce a linearly polarized beam, and is most easily achieved by passing unpolarized light through a polarizer, which allows waves of only one polarization to pass through. It should however, be noted that most common optical materials, like window glass, do not affect the state of polarization of the light, but some materials do (and is dealt with in the following chapters)—those that exhibit double refraction, or selective absorption or optical activity (chiral materials)—affect light differently depending on its polarization. Some of these effects are often used to make polarizing filters. It should also be noted that light becomes partially polarized when it reflects at an angle from a surface.

As is often necessary, one can define the states of polarization as illustrated in the images depicted in figure 1.10. Two different states of linear polarizations are illustrated in the first two diagrams—one that is horizontal (along the x-direction) and one at 45°, and each represent a distinct state of polarization (SOP). The linear

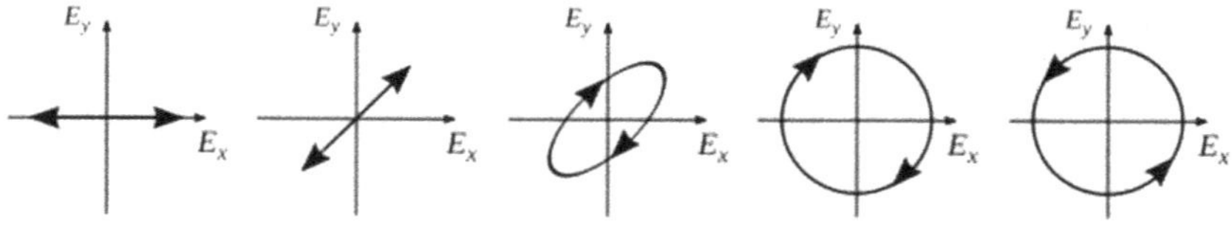

Figure 1.10. A schematic representation of the different states of polarizations of a light wave; the first two represent linear polarizations, the one in the middle, elliptical polarization and the other two, circular polarizations with opposite handedness. This Polarisation state image https://en.wikipedia.org/wiki/Polarization_(waves)# has been obtained by the authors from the Wikimedia website, where it is stated to have been released into the public domain. It is included within this article on that basis.

polarization at 45° can also be thought of as a linear combination of two linearly polarized waves: a horizontally linearly polarized wave (as in the leftmost image) and a vertically polarized wave with the same amplitude and phase.

Now then, should a phase shift be introduced between the horizontal and vertical polarization components, the resultant beam would generally be considered elliptically polarized, as illustrated in the middle figure. However, if the phase shift is exactly ±90° and the amplitudes are the same, then the resultant beam is said to be circularly polarized, as illustrated (fourth and fifth figures, but with opposite handedness). For the most part we shall be largely concerned with only linearly polarized light waves.

Thus far we have defined what a state of polarization represents; however, oftentimes it is necessary to provide a coordinate frame for describing the polarization of a propagating beam. This is necessary as sometimes it is necessary to describe things that happen to the polarization, for example rotation about the propagation direction. Hence, when one considers light that is propagating parallel to Earth's surface, it is convenient to use the terms 'horizontal' and 'vertical' polarizations to refer to the direction of the vibrating electric fields of the light.

One will often find in many textbooks another coordinate system used rather frequently and it is related to the plane of incidence. The plane of incidence is the plane that contains the propagation direction and the direction of reflection or refraction or scattering. If the electric field is parallel to this plane, then it is termed *p-like* (parallel) and if it is perpendicular then it is termed *s-like* (*senkrecht*, German of 'perpendicular'). Polarized light with its electric parallel to the plane of incidence is thus denoted *p-polarized*, and light whose electric field is normal to the plane of incidence is denoted *s-polarized*. Sometimes the *p-polarization* is also referred to as transverse-magnetic (TM) wave, and the *s-polarization* is also referred to as transverse-elective (TE) wave.

1.4.2 Linear polarizers

As already mentioned, a polarized light microscope is a compound microscope with two polarizers added in the optical path. When an unpolarized light wave is passed through an optical filter (a polarizer), what emerges is a polarized light beam with a specific polarization while blocking the other polarizations, thus producing linearly polarized light. Linear polarizers come in at least two varieties; absorptive polarizers, where the undesirable polarization states are absorbed by the filter, and beam-splitting polarizers, where the incident, unpolarized beam is split into two beams with orthogonal polarizations.

There are some crystals that possess what is referred to as *dichroism*, a tendency for preferential absorption of light that depends on the polarization directions, and hence, can be used as linear polarizers. A Polaroid polarizing filter was originally made of microscopic crystals of herapathite embedded in a polymer film, thus producing a linear polarizer. Currently the Polaroid polarizers are made by stretching a sheet of polyvinyl alcohol (PVA) with iodine doping. The process of stretching causes the PVA (polymer chains) to align in a particular direction and when doped with iodine, light polarized along the polymer chains is absorbed; light polarized perpendicularly to the polymer chains is transmitted, which then prescribes the polarization direction on the linear polarizer. Such polarizer sheets are relatively cheap to produce and hence, are used in modern microscopes.

For the sake of completeness in the description of polarization, we provide a short digression into polarizers created by reflection. Polarizers can also be made by having a light wave reflect at an interface between two transparent materials. This is because the reflectivity of light is different for light polarized in the plane of incidence (p-polarized) and light polarized perpendicular (s-polarized) to it. At a particular angle known as Brewster's angle, no p-polarized light is reflected from the surface, and hence, all reflected light must be s-polarized with its electric field vibrating perpendicular to the plane of incidence (horizontal polarization). Thus, a simple linear polarizer can be made by tilting a stack of glass plates at Brewster's angle (for visible light in air and glass, Brewster's angle is 57°) to the incident beam. For example, a stack of 10 plates (20 reflections), roughly about 3% of the s-polarized light is transmitted, and hence, the reflected beam is fully polarized, but is spread out and may not be a very useful configuration for a polarizer. This is schematically illustrated in the figure 1.11.

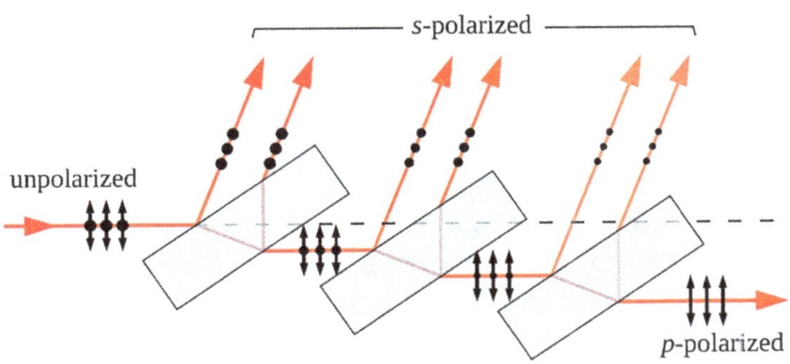

Figure 1.11. A stack of plates at Brewster's angle to a beam reflects off a fraction of the s-polarized light at each surface, leaving a p-polarized beam. Full polarization at Brewster's angle will require many more plates. The arrows indicate the direction of vibration of the electrical field and not the magnetic field which is perpendicular to them. This Brewster-polarizer image https://en.wikipedia.org/wiki/Polarization_(waves)# has been obtained by the authors from the Wikimedia website where it was made available by The Decryptor under a CC BY-SA 3.0 licence. It is included within this book on that basis. It is attributed to DrBob Bob Mellish.

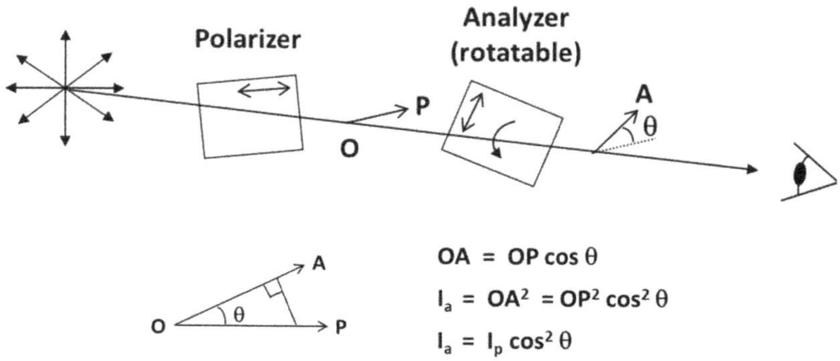

Figure 1.12. Illustration of an upolaraized light being sent through a polarizer and an analyzer, and the dependence of the intensity on the angle *q* between the two polarizers (Law of Malus).

In most modern polarized light microscopes, polarizers of the absorptive type are utilized, instead of the beam-splitter or birefringent type (Nicol's—see below). As already pointed out, the light that is transmitted through such a polarizer has its electric field vibrating in a direction normal to the alignment of the polymer chains, and hence, the direction normal to the polymer chains, is designated as the easy axis of the polarizer. If one were to place two polarizers one after another, the intensity of the light passing through depends on the angle between the easy axes of the two polarizers and the intensity is given by $I(\theta) = I_0 \cos^2 \theta$, where θ is the angle between the two polarizers. This is schematically illustrated in figure 1.12.

1.4.3 Doubly refracting materials

In the previous sections, we have described the state of polarization as well as how to create polarized light as well as what happens when a light beam is transmitted through two polarizers. We now begin with materials that are considered to be ordered and their interaction with light. There are a variety of materials that are transparent, examples being amorphous polymer melts, or normal glass found in windows, where the velocity of light passing through them is a constant irrespective of the direction of the passage of the beam; such substances are termed 'isotropic', and include many transparent amorphous substances such as glass and cubic crystals. There are a host of other materials, like liquid crystals and crystals (other than cubic crystals), however, for which the prior statement does not apply, and such materials are referred to as 'anisotropic'. Common examples of anisotropic materials include stressed glass, polymers that have crystallized, liquid crystals, as well as many types of crystals. When light passes through an anisotropic material, the light is split into two beams that are linearly polarized, with the polarizations of the two beams being mutually perpendicular. When an image or printed letters on a printed page are viewed through a calcite crystal, it is a remarkable sight to see each letter appear as double letters, as illustrated in figure 1.13 (right).

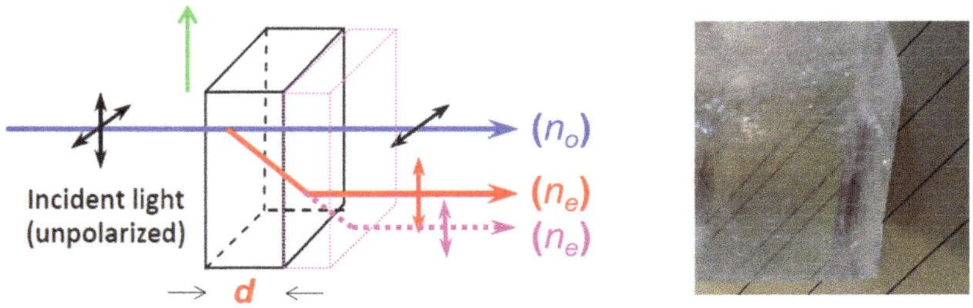

Figure 1.13. (Left) A schematic illustration of the path of a light beam through a doubly refracting crystal. The blue line indicates the 'ordinary ray' and the red line indicates the 'extraordinary ray'. The pink dotted lines depict an increase in thickness and the corresponding path length difference leading to the separation distance between the two rays. (Right) A quartz crystal laid upon a sheet of writing paper with the black lines showing the double refraction. The two outgoing beams are orthogonally polarized.

We shall deal with the simplest of these, the case of a uniaxial material where one of the rays obeys the normal laws of refraction, Snell's law, and has a constant velocity whatever its direction of propagation. This ray is called the '*ordinary ray*' and its refractive index is denoted by $\boldsymbol{n}_\mathbf{o}$. The other ray possesses a velocity that varies with the direction of propagation in the material, thus seemingly violating Snell's law, and hence it is known as the '*extraordinary ray*', and its refractive index is denoted by $\boldsymbol{n}_\mathbf{e}$. Thus, anisotropic materials possess two refractive indices, one for each of the two rays; while the refractive index is fixed for the ordinary ray, it is, however, variable for the extraordinary ray. Such substances are said to display 'double refraction', and the numerical difference between the refractive indices of the two rays is termed 'birefringence'. This is schematically illustrated in figure 1.13 (left).

We shall primarily be concerned with ordered materials, like polymeric spherulites or liquid crystals and the like. There are, of course, a variety of biological assemblies that have been studied using polarized light microscopy to understand the form and dynamics of cellular structures, where the assembly and disassembly dynamics of spindle microtubules have been revealed by utilizing polarized light microscopy. The interested reader/student is referred to various papers by Inoué and coworkers, and summarized in the book by Inoué [21], and we shall not discuss those here.

Materials displaying double refraction are easily studied by its effects on linearly polarized light which is allowed to propagate through a birefringent object, then has its resulting components recombined by transmission through a second polarizer with its easy axis normal to that of the incident polarization. As the two beams would have travelled different distances, thus inducing an '***optical path difference***', becoming greater as the birefringent object becomes thicker. As the two light beams or rays originate from the same light source, they are coherent, and thus there exists the possibility of either constructive or destructive interference, giving rise to a contrast when the two beams leave the material and are recombined at the second polarizer. It is this ability of birefringent objects that is used in polarized light microscopy. It should be noted that polarized light microscopy is often used not only to visualize ordered arrangements but is often used to make quantitative measurements of the birefringence, which is often used to prescribe the 'order' in liquid crystals, or the chain orientation in polymeric fibers, among other applications.

As indicated above, when a light beam is incident on a birefringent object (like calcite), the beam propagates in the object as two beams with orthogonal polarizations. This is schematically illustrated in figure 1.14 where the light beam is incident on a crystal slab cut in various directions. The blue rectangular slabs shown in panels a, b, c were cut from the crystal such that the ***optical axis*** in each of the panels is oriented differently (black double-headed arrow). In figure 1.14(a), a beam is incident on the crystal at an oblique angle to the optic axis, and is split into two rays, the ordinary and extraordinary rays, that travel different physical paths

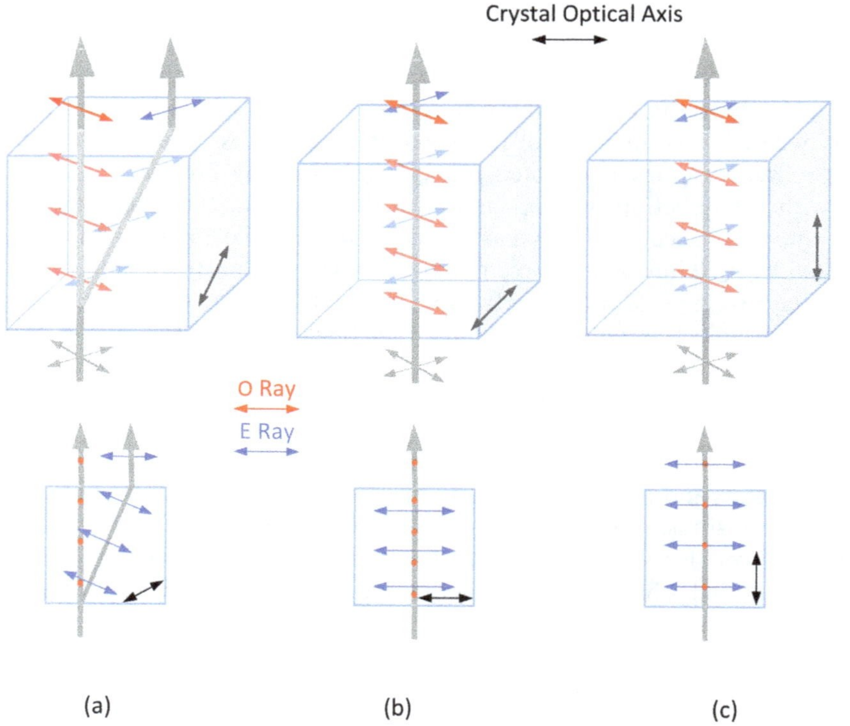

Figure 1.14. An incident ray is split into an ordinary (O) and an extraordinary (E) beam upon entering a birefringent crystal. The slabs shown in panels a, b, and c have been cut from parent crystals such that the optical axes are oriented in different direction with respect to the incoming light. The optic axis is represented by the double-headed black arrows. Both the 2D (lower) and 3D (upper) views are shown with the incident light being linearly polarized. Red and blue double-headed arrows indicate the planes of vibration of linearly polarized O and E rays, respectively. (a) An incident beam is obliquely incident to the optical axis and is split into two beams and the emergent O and E rays are linearly polarized but orthogonal to each other and exhibit a path difference. (b) An incident ray whose propagation axis is perpendicular to the optical axis is split into O and E rays, following the same trajectory through the crystal and do not diverge. Emergent rays can exhibit an optical path difference. (c) An incident ray propagating parallel to the optical axis is not split and behaves as an ordinary ray and experiences no path difference. Reproduced from [1] John Wiley & Sons.

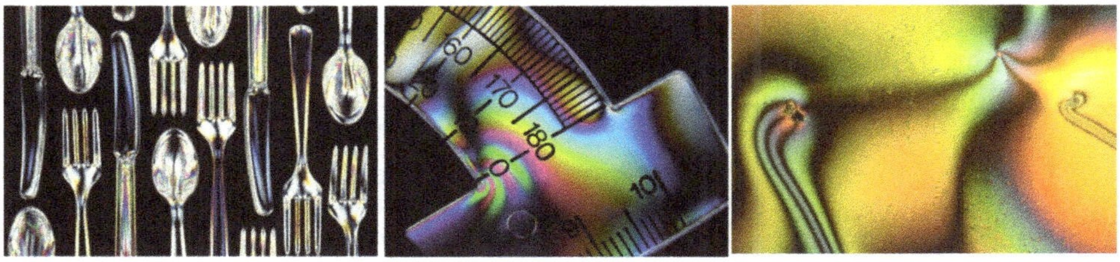

Figure 1.15. Photomicrographs of objects viewed under crossed polarizers: (left) common plastic spoons, knives and forks. This BiodegradablePlasticUtensils2 has been obtained by the authors from the Wikimedia website, where it is stated to have been released into the public domain. It is included within this book on that basis. (middle) A common protractor. This Plastic Protractor Polarized 05375 image has been obtained by the authors from the Wikimedia website where it was made available by Nevit Dilmen under a CC BY-SA 3.0 licence. It is included within this book on that basis. It is attributed to Nevit Dilmen. (right) A nematic liquid crystal.

through the crystal. The fact that the beam is split into is easily observed when placing a crystal of calcite on a printed page and looking down on the crystal while rotating a dichroic filter (a polarizer) held in front of the eye; the double letters become alternately visible and invisible as the filter is rotated through an angle of 90°. Incident light beams that are perpendicular to the optic axis split into ordinary (O) and extraordinary (E) rays, but the trajectories of these rays are coincident as illustrated in figure 1.14(b). At this unique angle of incidence, the O and E rays emerge at the same location on the crystal surface, but have different optical path lengths and therefore possess a phase shift. In figure 1.14(c) the optic axis is parallel to the propagation direction, and the beam travels in the material as an ordinary beam and experiences no path difference. Hence, when viewed with an analyzer that is orthogonal to the incident polarization, with the exception of (a), the beam is extinguished. However, in figure 1.14(c), when the sample is rotated, extinction will be maintained under crossed polarizers, thus behaving as if it were optically isotropic, like an amorphous glass.

A polarized light microscope is primarily a compound microscope which has two polarizers integrated into the microscope, and the polarizers' easy axes may be parallel or perpendicular to each other. When they are 90° to each other, it is referred to as crossed polarizers. The intensity of light that is transmitted by an anisotropic material between two crossed polarizers is dependent on the angle γ between the plane of polarization of the light beam and the optical axis of the birefringent medium as well as the optical path difference or the phase retardation $\boldsymbol{\delta}$ introduced by a sample of uniform thickness $\boldsymbol{d}$, and is given by the expression [8, 19]:

$$\text{Transmittance,} \quad T = \tfrac{1}{2}(1 - \cos 4\gamma)\ \sin^2 (\delta/2)$$

where $\delta = (2\pi\Delta nd/\lambda)$, Δn is the birefringence of the sample, and λ is the wavelength of light that is used. Often the angle γ is 45°, thus simplifying the expression for the transmission between two crossed polarizers. Thus, when birefringent materials are observed under crossed polarizers, they appear colored, due to the interference of the two beams propagating through the anisotropic medium, as illustrated by some common objects, as well as a liquid crystal.

These colors are easily interpreted and can be used to determine the birefringence of the object using the Michel–Levy chart. The Michel–Levy color chart is a reference for comparing the birefringence, thickness, and retardance (phase difference) of a given birefringent medium to its expected polarization colors. Figure 1.16 displays an excellent Michel–Levy chart (Zeiss). Approximately every 550 nm of retardance corresponds to a different order, with each order ending in red. It can be observed that the first-order colors are mostly grays and dull yellows, the second order is a brilliant array of all hues, and orders three and above show increasingly less saturated reds and greens until almost full achromaticity is observed around the tenth order. Hence, knowing the thickness of the object under observation, one can use the Michel–Levy chart to estimate the birefringence.

1.4.4 Polarization colors and the chromaticity diagram

As can be clearly seen, there are vivid colors when viewed under crossed polarizers. However, in order to discuss the colors observed, one has to pay attention to three necessary aspects that are not often discussed when attempting to understand the colors due to birefringence, and they are part of what is referred to as color science. However, we intend to demonstrate that since polarization colors have characteristic transmission wavelengths and are viewed by the *human eye* with a *specific light source*, they can be described with the simple tools of color science. Thus, a connection between the Michel–Levy color chart and the CIE chromaticity diagram will be established.

Consider a birefringent medium oriented under a microscope such that its optical axis is at a 45° angle with respect to the electric field vector of the plane-polarized light. Also, the propagating direction of the light beam is orthogonal to the medium's optical axis. The ensuing relationship that describes the intensity of transmitted light [$I\perp_{,45°}$] at a given wavelength is $I\perp_{,45°} = I_o \sin^2 \delta/2$, as already described. The Michel–Levy color chart is a reference for comparing the birefringence, thickness, and retardance (phase difference) of a given medium to its expected polarization colors.

The chromaticity diagram offers a mathematical description of color. The assumption is made that color has three requirements: *an object*, *an illuminant*, and a *color detector*. An object can be described by its spectral reflectance or spectral transmission curve, which maps the reflectance or transmission of light that interacts with the object at various wavelengths. The wavelengths of visible light range from around 400 nm to 700 nm, with 400 nm appearing violet and 700 nm corresponding to red. An illuminant is described by its spectral power distribution, which denotes the relative power of the illuminant through its gamut of wavelengths. Again, the range of wavelengths between 400 and 700 nm are of particular interest for the purposes of color science. Finally, the *sensation of color* requires a detector, such as the human eye. The CIE, or Commission International de l'Eclairage (International

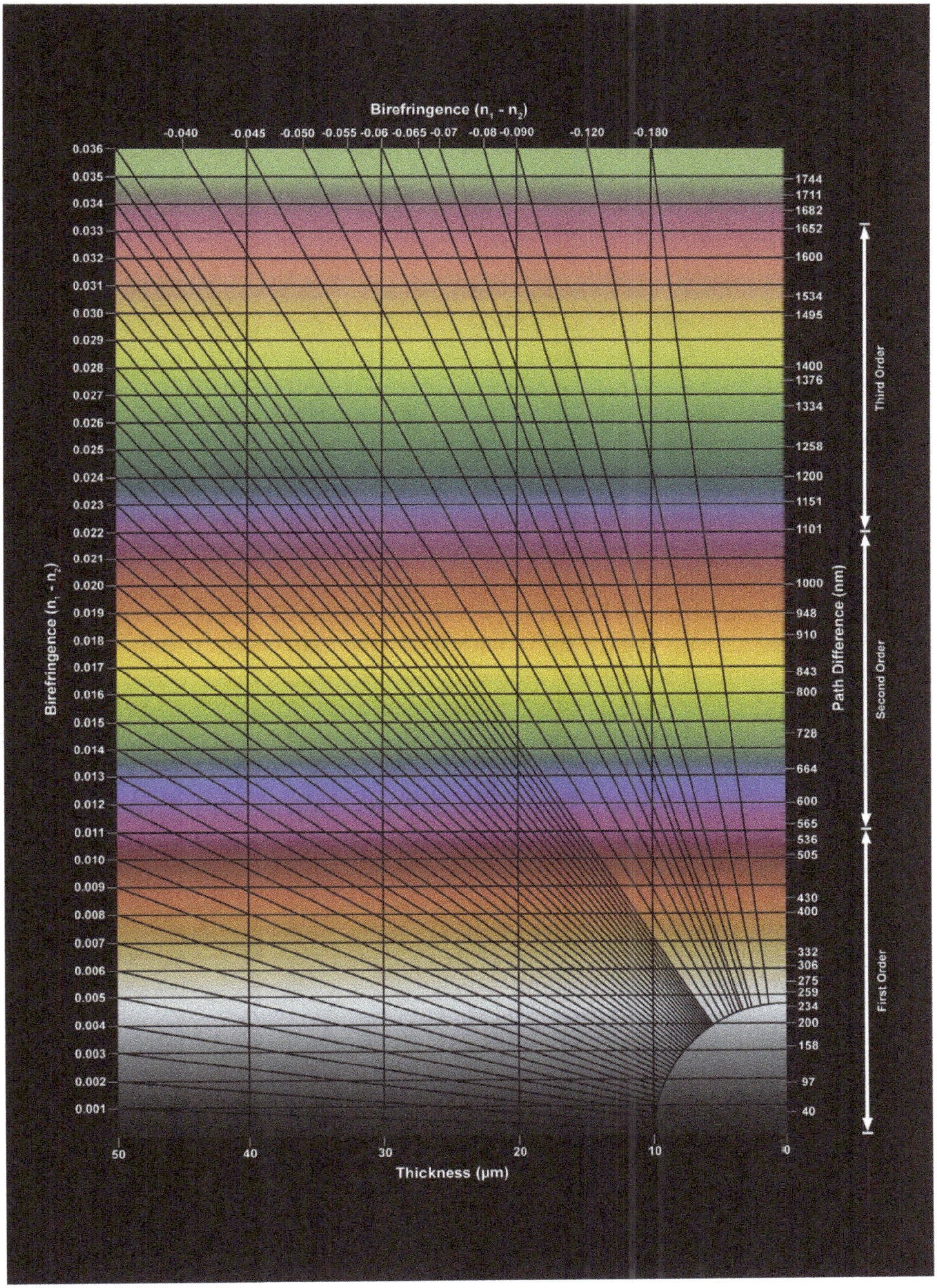

Figure 1.16. Michel–Levy chart illustrating the progression of colors with increasing retardation or optical path difference. Reproduced from Zeiss, with permission from Carl Zeiss, Inc., Thornwood, NY.

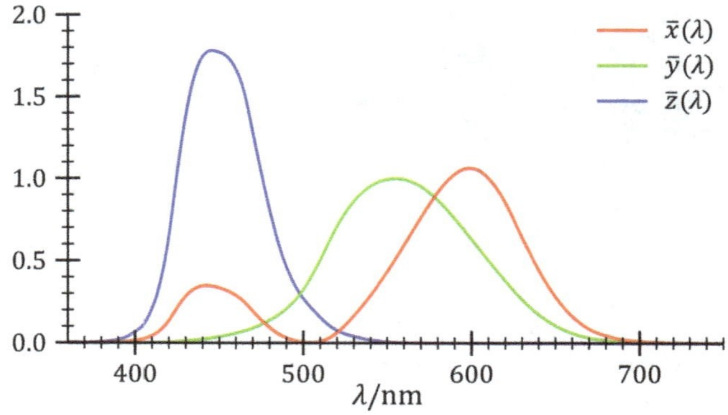

Figure 1.17. The CIE XYZ standard observer color matching functions. This CIE 1931 XYZ color matching functions image https://en.wikipedia.org/wiki/CIE_1931_color_space#Color_mixing_in_CIE_xyY_Creative CommonsAttribution-ShareAlike 4.0 license has been obtained by the authors from the Wikimedia website where it was made available by KdYpUvMgT under a CC BY-SA 4.0 licence. It is included within this book on that basis. It is attributed to KdYpUvMgT.

Commission on Illumination), decided on a standard set of color matching functions in 1931 so that color could be mathematically described [23]. This type of function can be generated by using a split field with one half illuminated by a single wavelength of light. The test for the observers is to match the other half of the field to the standard by adjusting the relative intensities of red, green, and blue light. In 1931, a large group of observers with normal color vision were asked to match every wavelength of visible light at a 2° viewing angle (so as to exclude any rod vision participation), and the CIE 1931 2° Standard Observer was born [24]. The Observer's perception of red, green, and blue is denoted by $\overline{x}$, $\overline{y}$, and $\overline{z}$, respectively. These color matching functions are displayed graphically in figure 1.17 [24].

By summing the transmission curve, power distribution, and color matching functions for a given object and light source over the visible spectrum of light, three so-called tristimulus values X, Y, and Z are obtained [23]:

$$X = k\int T(\lambda)\ \ P(\lambda)\overline{x}(\lambda)d\lambda$$

$$X = k\int T(\lambda)\ \ P(\lambda)\overline{y}(\lambda)d\lambda$$

$$X = k\int T(\lambda)\ \ P(\lambda)\overline{z}(\lambda)d\lambda$$

In the above equations, k is a constant which is determined by normalizing the relationships such that $Y = 100$ for a perfect white reflector which reflects 100% of all incident light. After normalization, CIE chromaticity coordinates x, y, and z are defined [24]:

$$x = X/(X + Y + Z)$$

$$y = Y/(X + Y + Z)$$

$$z = Z/(X + Y + Z)$$

Since the sum of the three chromaticity coordinates is most obviously equal to unity, only two of the coordinates are needed to describe a color. By convention, x and y are used, and these two values decide an object's color point on the CIE 1931 2° Observer chromaticity diagram. These two chromaticity coordinates aptly describe two of the three components of color, namely hue and saturation.

Hue is that property of color that distinguishes one small range of the visible spectrum from another, such as reds from greens. It is described by the various names we attribute to different colors. *Saturation* quantifies the purity of a specific hue as it departs from gray. A perfectly saturated color has only contributions from one specific wavelength of light, while a less saturated color has contributions from other colors. The third parameter of color is *luminance*, which is illustrated on a CIE color space with the tristimulus value Y. Luminance describes the lightness of a color, from black through the various grays to white (Committee, 1953). As previously mentioned, a Y value of 100 corresponds to a perfect white reflector, so white is the only color possible at $Y = 100$. As luminance decreases, more and more colors are attainable. It must be noted at this time that black, having a Y value of 0, does not occupy the entire lower plane of the color space. There is no such thing as a range of blacks; thus, black should be represented by a single point.

Although luminance is a vital part of color, the CIE (x, y, Y) color space can be difficult to grasp. Therefore, it is often more instructive to simply examine hue and purity by using the chromaticity diagram. A general guide to the arrangement of colors on the two-dimensional diagram is given in figure 1.18. The central area labeled CIE Ill C refers to the CIE standard illuminant approximating average daylight. Oftentimes, an unsaturated object color will take on the color of its illuminant, so this portion of the diagram could equally well be labeled gray. Any shade of gray, no matter how luminous, will have nearly equivalent tristimulus values X, Y, and Z. Thus, blacks, grays, and whites all lie around a color point of (1/3, 1/3). Points near the spectrum locus, or outer boundary, of the diagram are very saturated, which will be a point of interest with the polarization colors.

Using the transmission equation for a birefringent medium defined above, the spectral transmission curve for an object under a polarizing microscope can be calculated. The light source in most polarizing microscopes is some sort of halogen lamp, so a general spectral power distribution for a halogen light source can be assumed. Finally, the CIE 1931 color matching functions can be used as the color detector. Therefore, all the components needed for the chromaticity diagram are supplied, and it seems feasible that the color at a given retardation on the Michel–Levy chart will match well with its calculated position on the chromaticity diagram.

The relative powers for a tungsten–halogen lamp at a color temperature of 3100 K from 400 to 700 nm at 20 nm increments [24] and the 1931 color matching functions [24] were tabulated. After normalization of the product $P_{\mathrm{H}}\bar{y}$ to equal 1.0000, the final normalized products were calculated. Using equation for

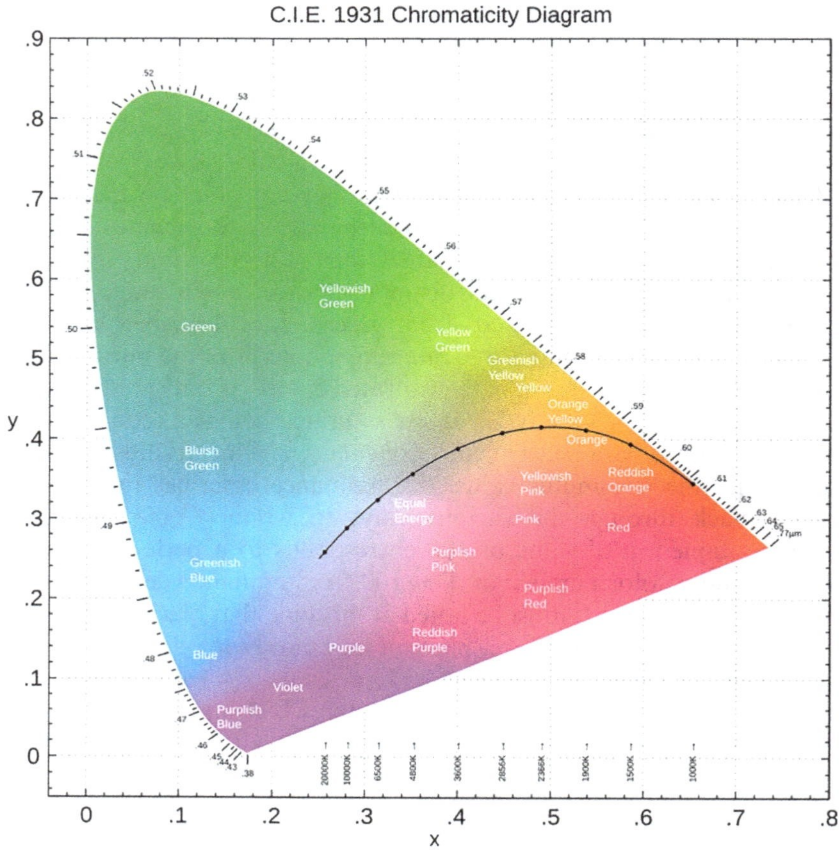

Figure 1.18. The CIE 1931 color space chromaticity diagram rendered in terms of the colors of lower saturation and value than those displayed in the diagram above that can be produced for printing purposes. This Chromaticity diagram full image has been obtained by the authors from the Wikimedia website where it was made available by Aboalbiss under a CC BY-SA 3.0 licence. It is included within this book on that basis. It is attributed to Paulschou.

transmission under crossed polarizers, relative transmission intensities were calculated in the visible light regime for retardations ranging from 550 to 2200 nm. Birefringence was assumed to be independent of wavelength for simplification. Figure 1.19 is a plot of the transmission for a birefringent material at 45° with various retardation values.

Finally, tristimulus values and chromaticity coordinates were calculated for every 25 nm of retardation. As mentioned before, 550–2200 nm corresponds to the second through fourth orders of the Michel–Levy chart, and these are of the greatest interest since they represent the most brilliant colors. Colors of the first order or greater than the fourth order are more or less achromatic and thus simply crowd the central area of the chromaticity diagram. The calculated color points are plotted on a CIE 1931 chromaticity diagram in figure 1.20.

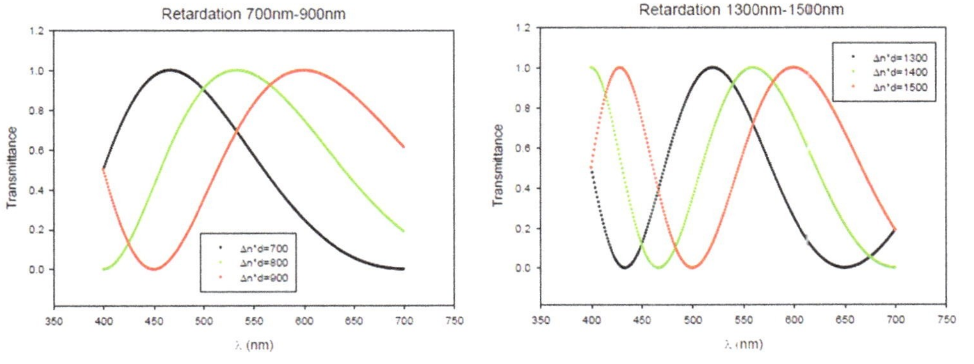

Figure 1.19. A plot of transmission of light through crossed polarizers for various retardation ($d\,\Delta n$), ranging from 700 to 1500 nm.

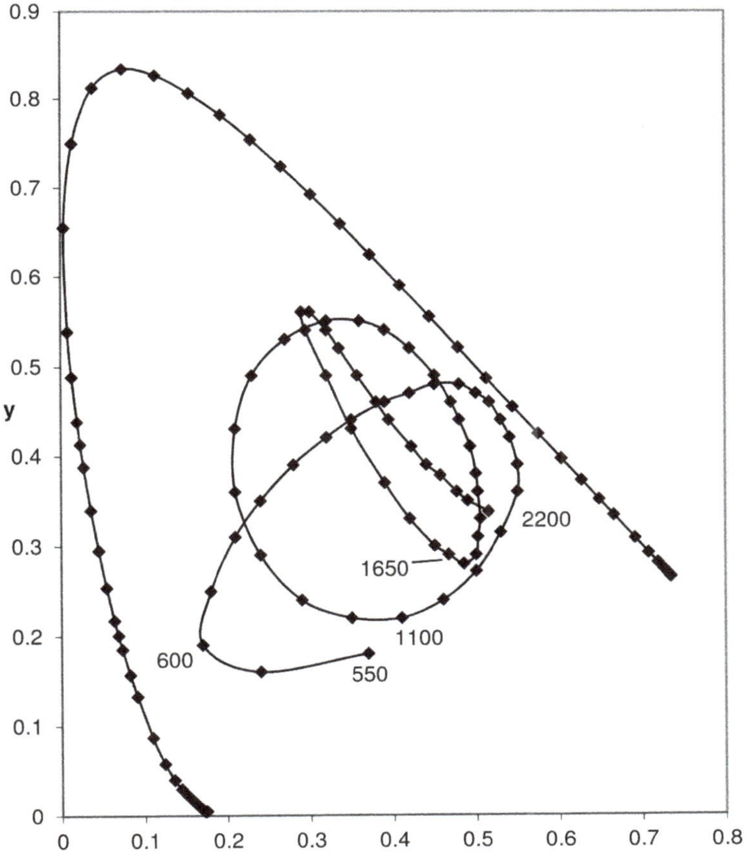

Figure 1.20. Chromaticity diagram illustrating chromaticity coordinates of polarization colors using a halogen light source for retardations of 550–2200 nm.

The second-order colors, ranging from 550 to 1100 nm, sample every hue rather well with a very smooth and slowly inflecting curve. Notice the relative brilliance of the purples and blues, and the extreme saturation of the yellows. The brilliance of these interference colors is one of the main reasons that scientists and commoners alike have been so fascinated by them over the years. The third-order polarization colors also sample all hues and have a smooth curve, but notice that the blues and oranges of this order are far less saturated than those of the second order. However, the third order has a much stronger representation of greens than the second order. The purples, blues, and yellows of the second order and the greens of the third order are similar to those which can be realized by an R, G, B color monitor, thus supporting the wide range of colors attainable by birefringent mediums. Finally, the fourth order begins to take the shape of a very slender curve merely oscillating between red and green. It has hardly any trace of magentas, blues, or cyans, and has a very small degree of saturation. Such is the trend with the higher orders of the Michel–Levy chart, which decompose to dull reds and greens, and eventually approach full achromaticity.

1.4.5 Conoscopy of nematic liquid crystals

As already discussed, a nematic liquid crystalline phase has uniaxial symmetry and possesses two refractive indices, the ordinary refractive index n_0 and the extraordinary refractive index n_e. Hence, optical characterization of these materials requires the measurement of two refractive indices. In the nematic phase, the measured birefringence Δn ($= n_e - n_o$) is directly related to the order parameter S defined as $S = (1/2)\,[(\langle 3\cos^2\theta\rangle - 1)]$. Therefore, measurement of Δn affords a simple way to characterize the order of the nematic phase, as a function of concentration in the case of a lyotropic liquid crystal and as a function of temperature in the case of a thermotropic liquid crystal.

1.4.6 Optical crystallography or conoscopy

In using this method for optical characterization, the nematic single crystal is illuminated by a highly convergent laser beam. In doing so, one is able to probe the sample at many different propagating angles of light simultaneously. This method, also referred to as optical crystallography, is important for quantitative analysis of the uniaxial material, be it a crystal or a liquid crystal. In order for the microscope to function as a conoscope (cono-converging; cone-shaped light wave for illumination), it is necessary to open the aperture diaphragm fully, so as to produce converging light (cone-shaped light) illuminating the uniaxial material. When an anisotropic material is viewed between crossed polarizers using highly convergent light, there will be produced in the back focal plane certain optical phenomena know as interference figures. Such interference figures allow us to determine the optical character of the crystal or anisotropic material. Based on the interference figures observed we will be able to determine whether the sample we are interrogating has two (uniaxial) or three refractive (biaxial) indicies. Of course, thus far in our discussions we have not utilized the Bertrand lens for any observations, other than when we wanted to observe the diffraction pattern from a grating dealing with the question 'how is an image formed?'. You may recall this from the discussion on Abbé's theory of image formation in a microscope as well as from the videos provided by Peter Evennett. The interference figures appear in the objective back focal plane (also referred to as the Fourier plane) where all of this takes place, normally not seen when using the microscope for routine observation (as an orthoscope), but can be seen by using a Bertrand lens.

The Bertrand lens (Amici–Bertrand lens) is an additional lens in the path of the light microscope, often inserted in the body tube above the objective and the analyser, but below the eyepiece. This lens is not focused on the specimen, but on the objective back focal plane, thus providing an enlarged view of the interference figures. The exact location of the back focal plane varies with the objective, and hence, the Bertrand lens often has focus control as well, in a full-fledged research microscope. For quantitative measurements on the interference figures in conjunction with the eyepiece micrometer, it is convenient to have a centering control on the Bertrand lens.

To understand how these interference figures are formed, it is useful to consider what happens to the converging beam as it enters the sample. One ray from the cone of light in air, in general, splits into two beams within the crystal. These two beams propagate with different velocities within the crystal, and hence would have traveled different path lengths, the two beams being polarized orthogonal to one another within the medium. All the emergent rays out the crystal are parallel to the incident ray, and may be collected by a lens or an objective. On passing through an analyser the rays are brought to interference in the focal plane of the lens or objective. The interference figure so formed represents a distribution of light which can be described by curves of equal intensity. Every point in the focal plane (back focal plane) corresponds to a direction of parallel rays entering and leaving the crystal, and therefore one must consider the phase difference δ with this direction.

This method is especially useful for samples with large birefringence, since orthoscopic polarized light microscopy will not reveal small distortions of the optical axis within the sample plane, due to the so-called adiabatic limit, where the polarization of the incident beam is rotated in the sample [25]. By contrast, such small distortions of the director field are revealed by rotation of the conoscopic interference figures about its center of symmetry as the director field is rotated in the sample plane, or by translation of the pattern if the director field tilts out of the sample plane. Figure 1.21 is a schematic diagram of an optical microscope for conoscopic measurements. The conoscopic figures are observed in the rear focal plane of the objective (see figures 1.6 and 1.21).

For uniaxially symmetric materials, with the optical axis normal to the axis of the microscope, the conoscopic interference figures are hyperbola-like principal isochromates (curves of zero intensity) placed along and orthogonal to the optical axis (or director) of the material [8, 26]. For a material with positive birefringence the retardation $\delta = 2m\pi + \delta'$ for light propagating normal to the sample plane, m being an integer and $0 < \delta' < 2\pi$; the successive fringes along the optic axis have retardation $2m\pi$, $2(m-1)\pi, \ldots, 2\pi$ on moving away from the center, and those along the orthogonal axis have retardation $2(m + 1)\pi$, $2(m + 2)\pi\ldots$ on moving outward.

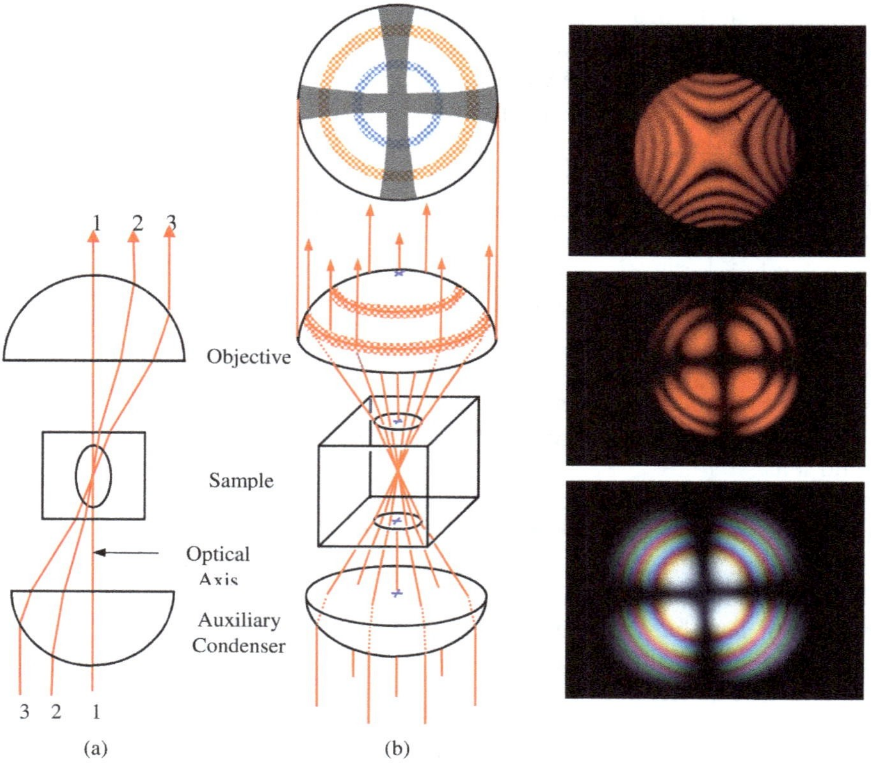

Figure 1.21. The figure schematically illustrates the formation of interference figures in the back focal plane of the objective when crystals are illuminated with a convergent beam of light. In (a) it is assumed that the optic axis of the crystal or the nematic monodomain is parallel to the propagation direction of the beam. In that case beams following path labeled '1' experience zero retardation while paths 2 and 3 produce progressively higher retardations. Because of the symmetry the interference pattern is a set of concentric circles as shown in the bottom two images, one in white light (the bottom most) and the one in the middle is exposed to a He–Ne source (633 nm). On the top is shown the interference figure when the optic is orthogonal to the propagation direction of the light beam, and presents a set of Hyperboli. Reproduced from [1] John Wiley & Sons.

For many liquid crystals used m can be large, say, $m = 20$ or more, depending on the liquid crystal that is being used and the cell thickness. The birefringence of the samples can be determined by measuring the distance between successive fringes that are placed normal to the director or the optical axis. Experimentally, the distance between the fringes can be measured using a calibrated eyepiece on the microscope or by using a photograph of the interference figures. A set of interference figures are shown in figure 1.21 for a nematic liquid crystal with the optic axis in the plane of the stage of the microscope and the optic axis normal to the microscope stage. An interference figure for a quartz crystal with its optic axis parallel to the optical path of the microscope is also shown in the figure. Figure 1.21(a) shows the interference figure for a monodomain of a lyotropic chromonic liquid crystal with its optical axis in the plane of the microscope stage where the center of symmetry of the interference figure is centered along the optical axis of the microscope. Should the optic axis be tilted with respect to the optical axis of the microscope, the center of symmetry will no longer at the center of the field of view, thus providing the valuable information that the optical axis of the system is tilted, which is of importance for understanding the evolution of the optic axis in response to a shear flow, for example. Should the optic axis be rotated in the plane of the sample, the conoscopic figures will also rotate about the optical axis of the microscope. The rotation of the interference figure is related to the rotation of the optical axis of the crystal, as is discussed below. This is the principle that is used to measure the twist viscosity of the nematic liquid crystals [25].

For a continuous rotation of the director field, the rotation Φ_{ROT} of the interference figures is given by

$$\tan 2\Phi_{ROT} = \langle \sin 2\phi \rangle / \langle \cos 2\phi \rangle$$

where ϕ is the twist angle, and the angular brackets denote the average of twist $\phi(z)$ through the sample thickness z $(-d/2 \leqslant z \leqslant d/2)$. For a material at equilibrium in a field imposing a continuous twist distortion, with maximum twist $\phi(0) = \phi_{max}$ in the midplane of the sample and zero twist $\phi\,(d/2) = \phi(-d/2) = 0$ at the surfaces, Φ_{ROT} may be expressed as

$$\tan \Phi_{ROT} = \frac{2\sin(\varphi_{max})}{2E\left(\frac{\pi}{2}, \sin\varphi_{max}\right) - F\left(\frac{\pi}{2}, \sin\varphi_{max}\right)} \tag{1.1}$$

where E and F are the complete elliptic integrals of the first and the second kinds, respectively. In this analysis, $\phi(z)$ is essentially given by $\phi_{max}\cos(\pi z/d)$ for small ϕ_{max}, though the rigorous expression is more complicated.

For the range of ϕ_{max} of interest here, calculations using the above equation show that ϕ_{max} may be approximated from the measured Φ_{ROT} by the simple expression $\phi_{max} = 1.57\Phi_{ROT}/(1 + 0.14\,\Phi_{ROT}^2)$ within 3% error (with Φ_{ROT} in radians). By comparison, a bulk rotation of the director field, such that $\phi(z) = \phi_{UNF}$ at all z, gives $\Phi_{ROT} = \Phi_{UNF}$ for the range of retardations often encountered. For the sake of completeness, the translation Δ of the center of symmetry of the fringe pattern from the optical axis of the microscope is related to an average of the tilt $\theta\,(z)$ of the director out of the sample plane. For a uniform tilt ($\theta(z) = \theta_{UNF}$), $(2\Delta/D)NA = n_1$ $(\sin\theta_{UNF})$, where D is the diameter of the field of view, NA is the numerical aperture of the objective, and n_1 is the average refractive index.

Thus, it should be apparent that the application of an optical crystallographic technique, conoscopy, to liquid crystals can lend itself to the measurements of static properties of the medium, i.e. measurement of the birefringence of the liquid crystal or the anisotropic crystal, and the two individual refractive indices of the medium. In some cases, these can be measured as a function of wavelength as well to understand the variation in the refractive index as a function of wavelength [26]. In other words, the dispersion of the system can be experimentally measured.

Consider an interference figure centered on the optic axis of the microscope. A light beam traveling normal to the uniaxial crystal experiences the maximum Δn. The center of the interference figure is illuminated by such a beam. Suppose that the retardation is $m\pi + \delta$, at the center. A beam traveling parallel to the plane (the plane is defined by the optic axis of the microscope and the optic axis of the crystal) containing the optic axis of the liquid crystal experiences a smaller Δn, Δn_{eff}, than a beam traveling in the plane normal to that containing the optic axis, which at all angles experiences Δn. In the former case, Δn_{eff} is given by n_{eff}–n_{o}, where $n_{\text{eff}}^{-2} = n_{\text{o}}^{-2} \cos^2\theta + n_{\text{e}}^{-2} \sin^2\theta$, with θ the angle between the optic axis of the liquid crystal and the propagating beam in a plane containing the optic axis. For a beam traveling at an angle q, the retardation can be approximated by $\Delta n \sin\theta$. Thus, in general, the fringes along the optic axis experience a smaller retardation and will have successively lower order, say, $m\pi$, $(m-1)\pi$, etc, compared with the center of the interference figure, as discussed above. All beams traveling at different angles in a plane normal (the plane is again defined by the optic axis of the microscope and the normal to the optic axis of the liquid crystal) to that containing the optic axis of the liquid crystal experience the same Δn. Knowing this helps in evaluating the birefringence of the uniaxial crystal.

The birefringence of a uniaxial crystal can be determined from the interference figures. As discussed above, beams traveling normal to the plane containing the optic axis will see the Δn of the crystal. Hence, if one has more than one set of fringes normal to the optic axis (there are four sets in the image shown), these can be used to determine Δn. The retardation between two sets of fringes is equal to 2π. Thus, the difference in the retardation between the two sets of fringes can be written as

$$\delta_2 - \delta_1 = (2\pi d\Delta n/\lambda)\left[\frac{1}{\cos\theta_2} - \frac{1}{\cos\theta_1}\right]$$

Knowing the propagating angles within the uniaxial crystal, θ_1 and θ_2, which can be measured from the distances in the back focal plane, it is then trivial to calculate the birefringence of the uniaxial crystal no matter how many orders of retardation are present in the material of interest. Clearly, a similar strategy can be used to determine Δn_{eff}, and hence, one can easily solve for the independent refractive indices, n_{e} and n_{o} [26].

References

[1] Srinivasarao M 2022 New stuff in old places: light microscopy of polymers *Macromolecular Engineering: From Precise Synthesis to Macroscopic Materials and Applications* (Wiley) pp 1–82

[2] Valberg A 2005 *Light Vision Color* (Chichester: Wiley)

[3] Krebs H A 1970 History of tricarboxylic acid cycle *Perspect. Biol. Med.* **14** 154

[4] Ford B J 2009 Did physics matter to the pioneers of microscopy *Advances in Imaging and Electron Physics* vol 158 P W Hawkes (San Diego: Elsevier) p 27

[5] Ford B J 1992 Brown observations confirmed *Nature* **359** 265

[6] Carpenter W B 1891 *The Microscope and its Revelations* (Philadelphia: Blakiston)

[7] Turner G L'E 1979 *The Great Age of the Microscope* (New York: IOP Publishing) p 375
[8] Born M and Wolf E 1980 *Principles of Optics* 6th edn (New York: Pergamon)
[9] Betzig E 2015 Nobel lecture: single molecules, cells, and super-resolution optics *Rev. Mod. Phys.* **87** 16
[10] Lipson S G, Lipson H and Tannhauser D S 1995 *Optical Physics* (Cambridge: Cambridge University Press)
[11] Betzig E and Trautman J K 1992 Near-field optics: microscopy, spectroscopy, and surface modification beyond the diffraction limit *Science* **257** 189–95
[12] Masters B R 2020 Superresolution optical microscopy *The Quest for Enhanced Resolution and Contrast* Springer Series in Optical Sciences Series in Optical SciencesSpringer vol 399 (Cham: Springer)
[13] Kohler H 1981 On Abbé's theory of image-formation in the microscope *Opt. Acta* **28** 1691–701
[14] Porter A B 1906 On the diffraction theory of microscopic vision *Philos. Mag.* **11** 154–66
[15] Goodman J W 1996 Introduction to Fourier optics *Fourier Optics* vol xviii (New York: McGraw-Hill) p 441
[16] Di Francia G T 1956 On resolving power and information *J. Opt. Soc. Am.* **46** 72
[17] Rayleigh L 1896 On the theory of optical images, with special reference to the microscope *Philos. Mag.* **42** 167–95
[18] Dunst S and Tomancak P 2019 Imaging flies by fluorescence microscopy: principles, technologies, and applications *Genetics* **211** 15–34
[19] Hecht E and Zajac A 1979 *Optics* (Reading, MA: Addison-Wesley) p 551
[20] Francon M 1961 *Progress in Microscopy* (New York: Row, Peterson, and Company) p 289
[21] Inoué S 1997 Video microscopy: the fundamentals *The Language of Science* 2nd edn ed K R Spring (New York: Plenum)
[22] Pearson H 2007 The good, the bad and the ugly *Nature* **447** 138–40
[23] Billmeyer F W and Saltzman M 1981 *Principles of Color Technology* 2nd edn (New York: Wiley) pp 58–65
[24] Wyszecki G and Stiles W S 1982 *Color Science: Concepts and Methods, Quantitative Data and Formulas* 2nd edn (New York: Wiley) p 19
[25] de Gennes P G and Prost J 1993 *The Physics of Liquid Crystals* 2nd edn (Oxford: Clarendon)
[26] Mattoussi H *et al* 1992 Refractive-indexes dispersion and order of lyotropic liquid-crystal polymers *Macromolecules* **25** 2860–8

Part II

Seven experiments (open-ended)

This part deals with seven experiments that are designed to give you a good understanding the subject matter dealing with light microscopy of soft materials —also is useful for understanding the material that is taught in introductory undergraduate materials science and engineering courses throughout the United States—perhaps around the world dealing with materials science and engineering.

IOP Publishing

Through a Modern Microscope
Optical experiments on soft matter
Mohan Srinivasarao, Jung Ok Park and Karthik Nayani

Chapter 2

Experiment 1: Getting to know your optical microscope (DM 750P) and its basic software

2.1 Getting started

1) Computer login: Use your GT ID and password;
2) Create your own folder in D Drive;
3) Open LAS (Leica Applications Suite) by double clicking;
4) Find your own folder in LAS → Select it → Click the default folder sign → The folder with a red dot will be your default folder;
5) Choose the image file format: JPEG.

It should be noted, while we have used a Leica for illustration, this description is applicable to most polarized light microscopes, irrespective of the manufacturer. All Leica images courtesy of Leica Microsystems.

doi:10.1088/978-0-7503-1431-2ch2

2.2 Key structure of the Leica DM 750P (transmission mode)

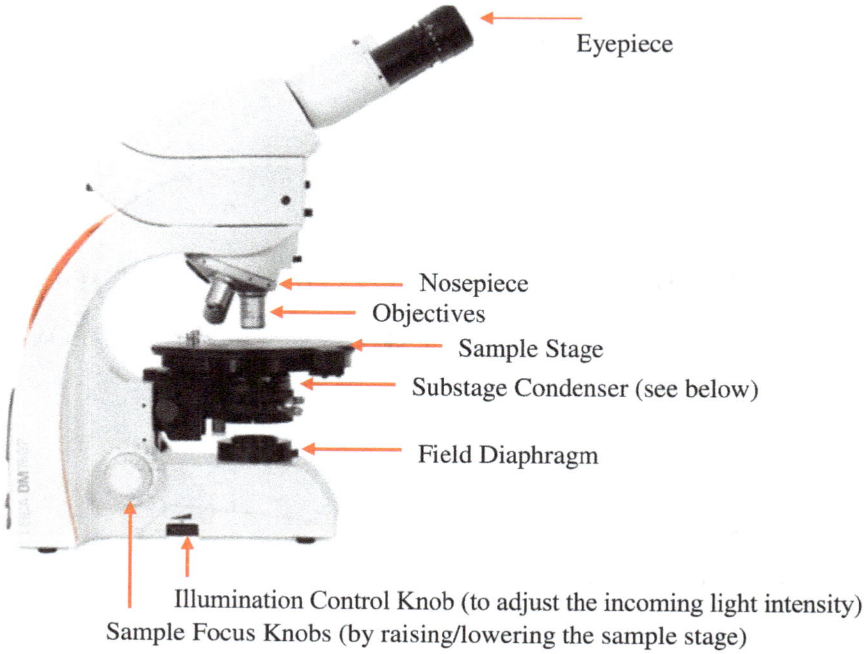

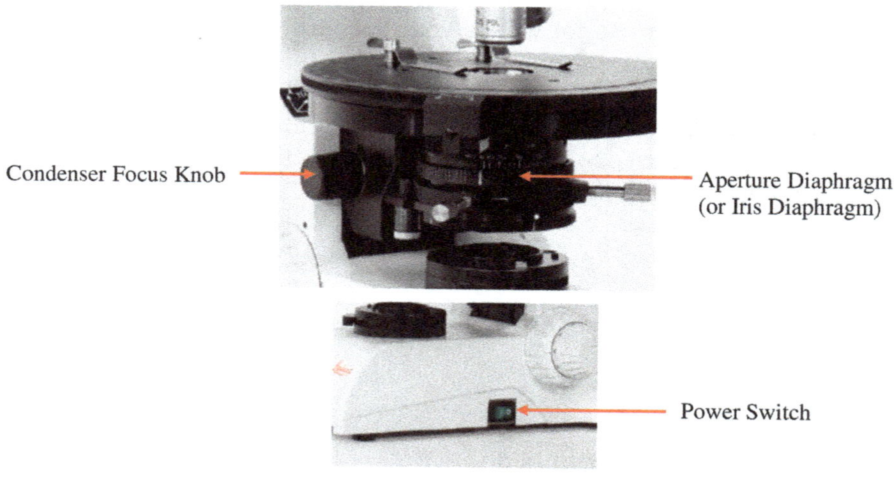

2.3 Focusing

1) Place a specimen on the stage.
2) Rotate the 10× objective into working position.

3) Raise the microscope stage using the coarse adjustment knob until you reach its positive stop or the specimen almost touches the objective.

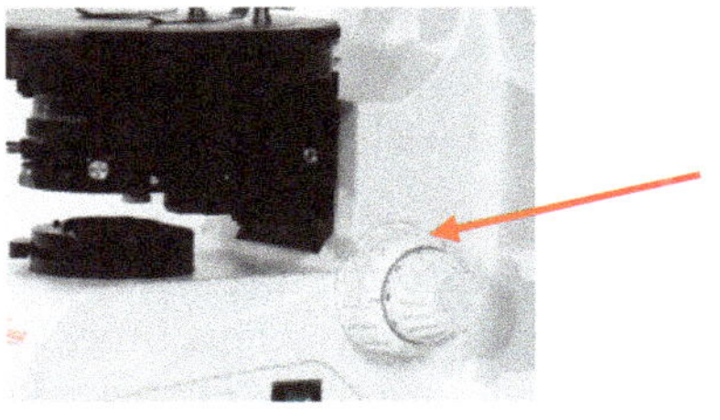

4) Look into the eyepieces and adjust the illumination intensity so that it is comfortable for viewing.
5) Using the fine focus adjustment knob, bring the specimen into sharp focus.

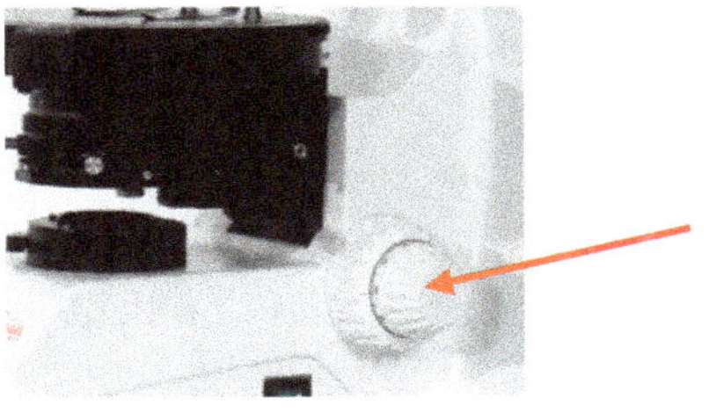

2.4 Viewing tube adjustment

1) Fold or unfold the eyepieces until you see one illuminated circle. To correctly set the eyepieces, be sure the crosshair eyepiece is seated securely in the slot of the right eyetube.

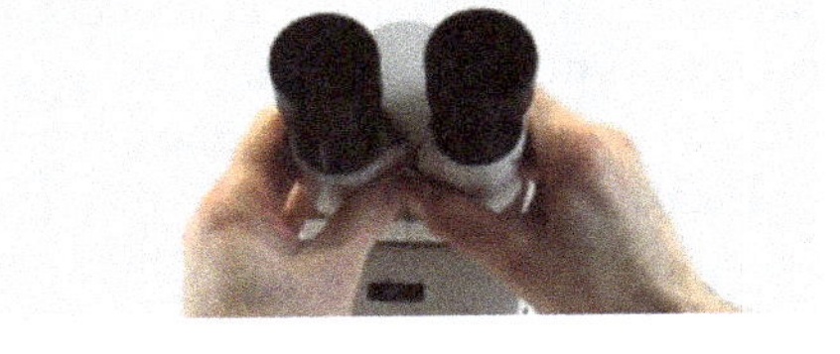

2) Focus the crosshair reticle by gripping the knurl and rotating the top of the eyepiece until the reticle is in clear focus.
3) While looking through only the crosshair eyepiece, bring the specimen into focus with the fine focus of the microscope.

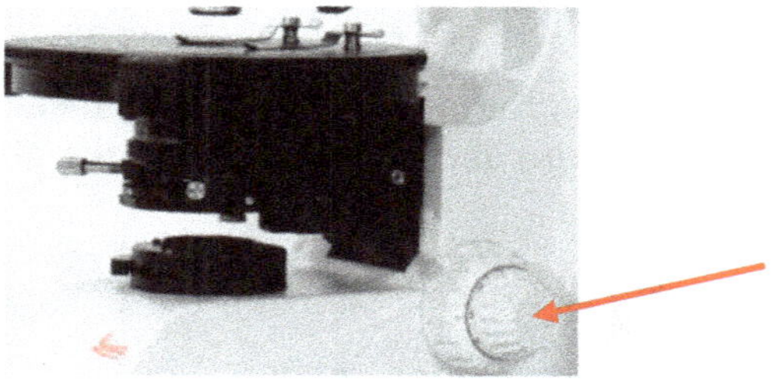

4) Now switch to look only through the other eyepiece (focusing eyepiece without crosshair) with the other eye. This time, focus the specimen by using the focusing capability in the focusing eyepiece. Do not raise or lower the stage to focus!
5) Grip the knurl on the focusing eyepiece with one hand and rotate the top of the eyepiece with the other hand until the specimen is in focus with this eye and this focusing eyepiece. This corrects for any vision differences between your right eye and left eye.
6) Now switch to a high magnification objective (not an oil objective) and focus the microscope while viewing with both eyes.

2.5 Köhler illumination set up

We will try and understand the principle of Köhler illumination by understanding figure 2.1 [1]. Figure 2.1 shows the ray optics in a light microscope which has been set for Köhler illumination and also shows the locations of the different lenses, field planes and aperture planes.

To the left of figure 2.1 is what are called the conjugate field planes of an optical microscope. The word 'conjugate' is used because in a well aligned microscope, all conjugate planes are simultaneously in focus. The conjugate field planes are what you see when you see through the eyepiece of an optical microscope. Clearly the four conjugate field planes are the Retina, intermediate image plane, object plane and the field (stop) diaphragm. So, if you see a dust particle through the eyepiece, it could in principle be on any of these four planes, because they are all in focus!

The right-hand side has what are called the conjugate aperture planes. The Iris diaphragm, back focal objective plan, front focal plane of the condenser and the lamp filament are the four conjugate aperture planes. To view the aperture planes, we need what is called the Bertrand lens. If you had observed the aperture planes in figure 2.1 carefully, you would have realized that the aperture planes can also be viewed by simply removing the eyepiece.

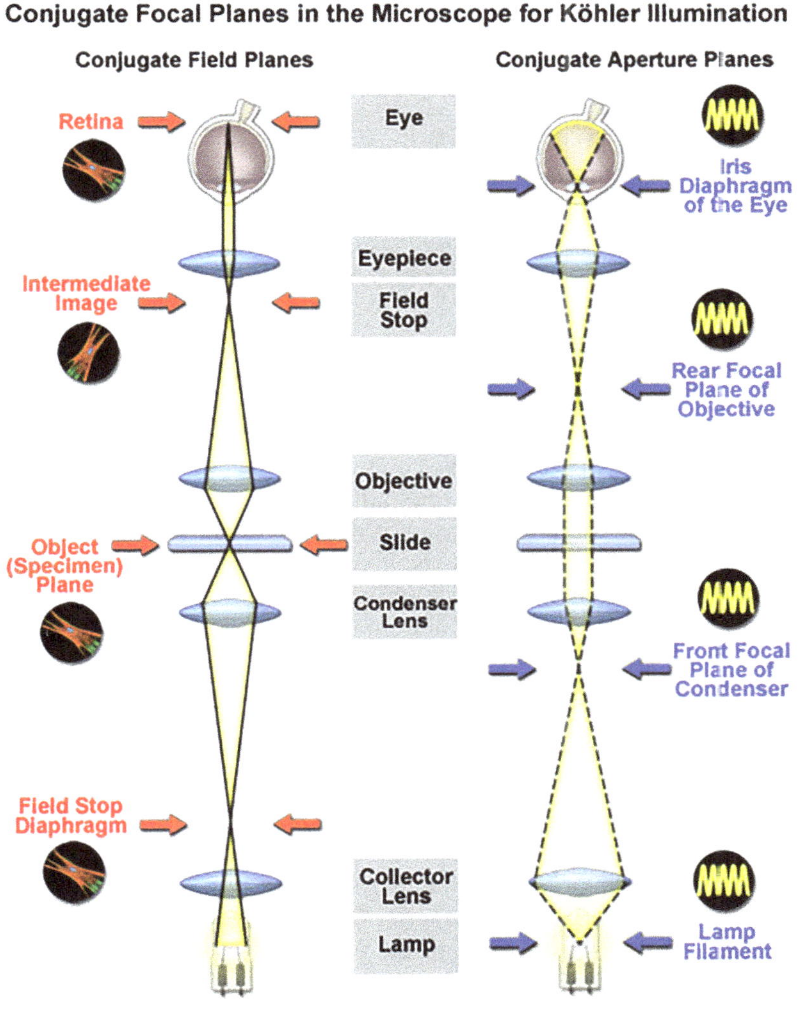

Figure 2.1. Illustrates the light paths in an optical microscope that is set up for Köhler illumination and the conjugate focal planes. Image credit: MolecularExpressions.com at Florida State University.

The crucial thing to understand from these two ray diagrams is the fact that the object plane is conjugate to the field (stop) diaphragm and the lamp filament is conjugate to the front aperture plane of the condenser (also known as the aperture diaphragm). This becomes important because the aperture and field diaphragms can be adjusted by the user for optimal illumination. It is not hard to see from the ray optics that the field diaphragm sets the field of view, while the aperture diaphragm controls the amount of light entering the microscope. So with this basic understanding, which diaphragm do you think affects the contrast and which diaphragm controls the resolution of an image? Understanding figure 2.1 well will help you understand the principle behind Köhler illumination thoroughly. Here is another small thought experiment for the same purpose:

To control the amount of light entering the light microscope, we could either:

> *(1) Increase the intensity of the bulb and narrow down the aperture diaphragm, or (2) work at a lower intensity of the bulb and open up the aperture diaphragm more. Which do you think is the better way to operate? You can think about this just by looking at figure 2.1. (Do so before you read the steps for Köhler illumination, which has the answers to both the above questions!!)*

Before looking at the steps for setting up an optical microscope for Köhler illumination, here are *the main purposes Köhler illumination* serves:

a) **Even illumination in the object plane**: This can again be understood when one looks at figure 2.1 carefully. We can see that when the field (stop) diaphragm and aperture diaphragms are optimally opened, the ray diagram of the field conjugate planes shows how each point is illuminated by the whole of the lamp, and the ray diagram of the aperture conjugate planes shows how the whole of the specimen can the illuminated by each point on the lamp. This is the principle behind the even illumination using this technique.

b) **Maximum resolution and contrast**: This would seem to be more intuitive, that we need to have the field and aperture conjugate planes in focus to have maximum resolution. The interplay between resolution and contrast, however, is a little more intricate and we will get a better picture hopefully, when we look at the steps of setting up Köhler illumination.

The procedure for the Köhler illumination for our Leica 750P microscope

1) Close the (Köhler) field diaphragm on the base of the microscope so that its iris leaves are present within the field of view when looking into the eyepieces.

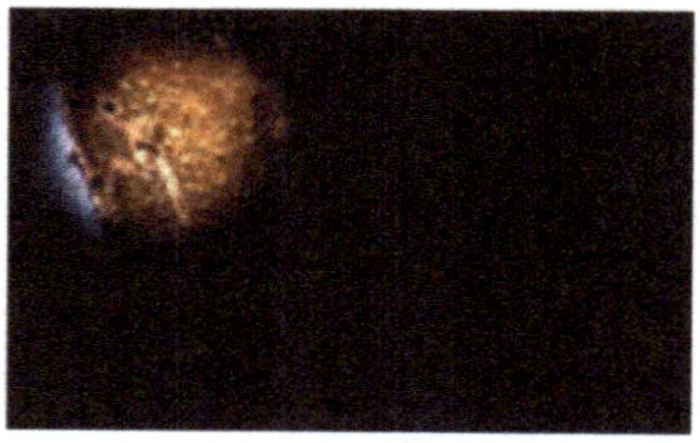

2) Use the condenser focus knob on the left-hand side of the substage to bring the leaves of the field diaphragm into sharp focus.

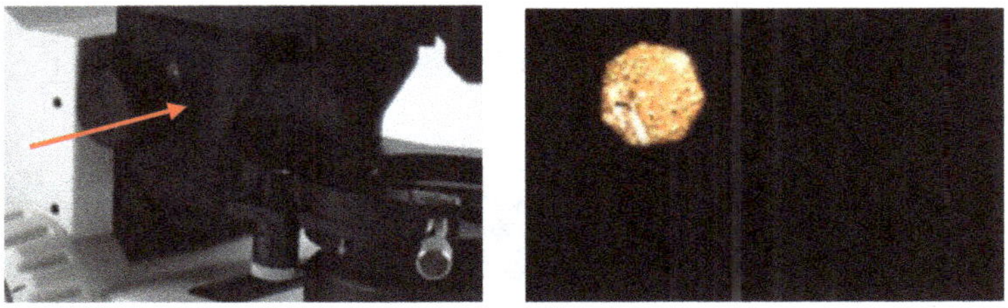

3) Turn the condenser centering thumbscrews simultaneously to center the image of the field diaphragm in the field of view.

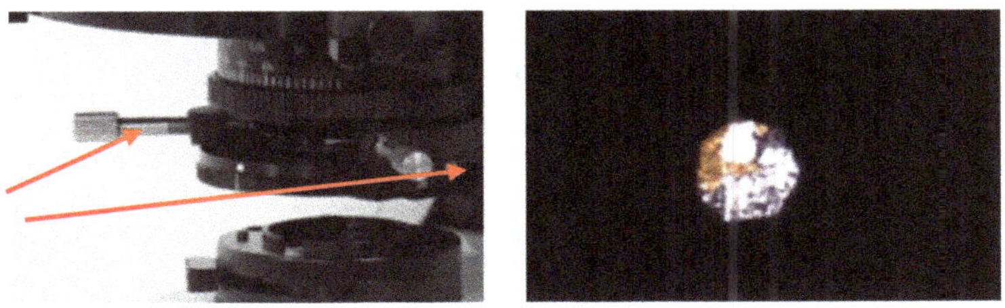

4) Open the field diaphragm until the iris leaves disappear just beyond the field of view.

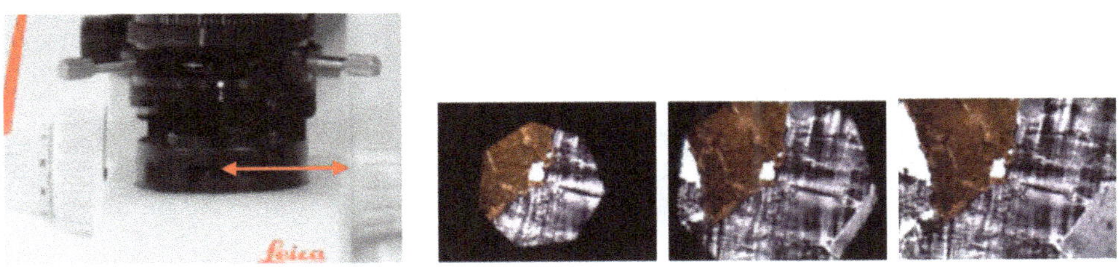

2.6 Setting up polarized light

The polarizer is located under the substage condenser in a swing in/out mount. Swing it into place and be sure it 'clicks' into place.

1) Loosen the polarizer locking thumbscrew, which is located on the right-hand side of the polarizer.

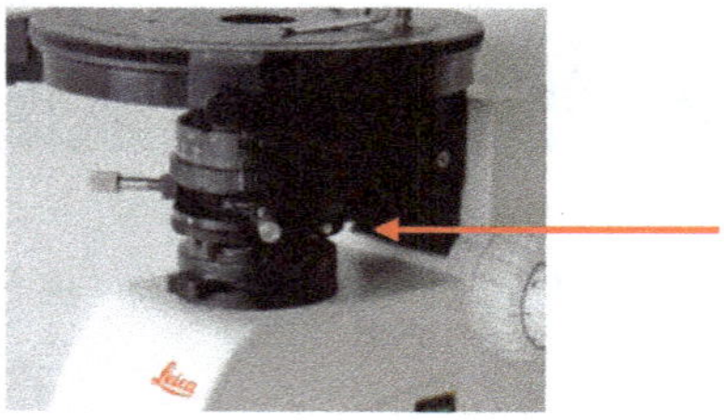

2) Rotate the polarizer to the zero position. The Analyzer is located in an Analyzer Bertrand Lens module.

Slide the Analyzer into position.

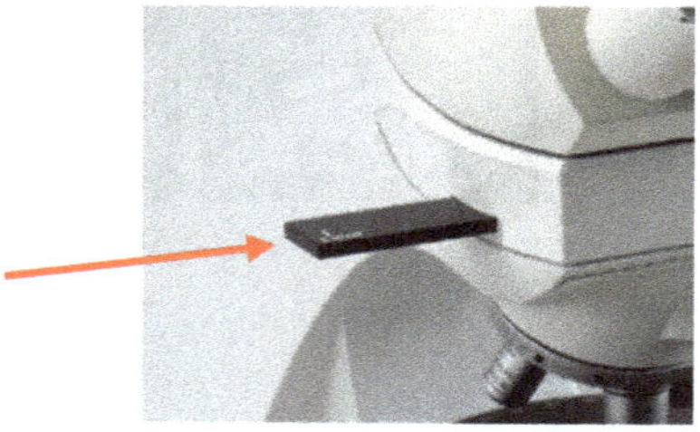

3) If you have an Analyzer/Bertrand Lens module, swing in the Analyzer by flipping the lower rocker of the module to the right (the 'A' position/ Analyzer in), while ensuring the upper rocker of the Analyzer/Bertrand Lens Module is flipped to the left (the 'O' position/Bertrand Lens out).

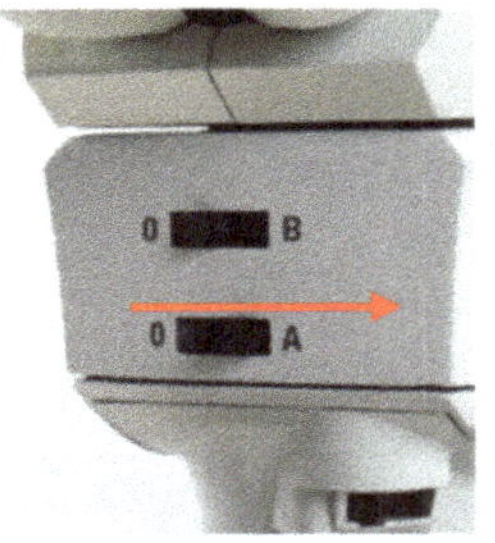

4) Increase the illumination intensity slightly. You will notice while looking into the microscope with the Polarizer and Analyzer in place and no specimen, that the field will look dark. If not, rotate the polarizer slightly until you have the darkest field of view.

2.7 Bertrand lens operation

If you have an Analyzer/Bertrand Lens module, you can view a conoscopic image through the eyepieces. Bertrand lens is also used to see the diffraction pattern of your sample, which will be the topic of lab 3 and 4.

1) Bring the portion of the specimen you would like to view in conoscopy (with the Bertrand Lens) to the center of the field.
2) Swing in the Bertrand Lens by flipping the upper rocker of the Analyzer/ Bertrand Lens module to the 'B' position. Be sure the Polarizer and Analyzer are also in position. You can now view the conoscopic image through the eyepieces.

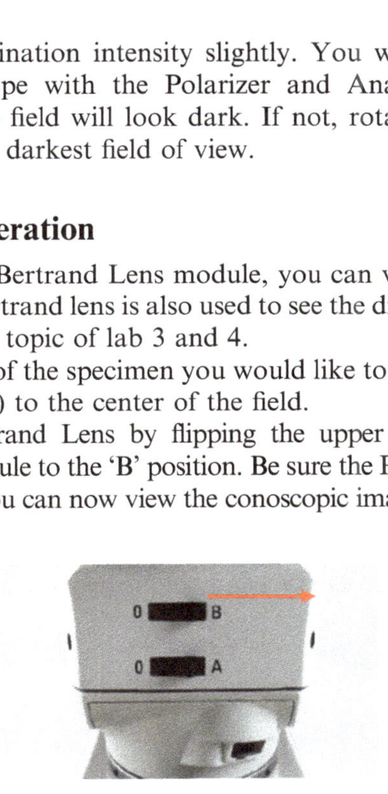

3) If necessary, you can fine tune the Bertrand Lens centration with the tools provided.

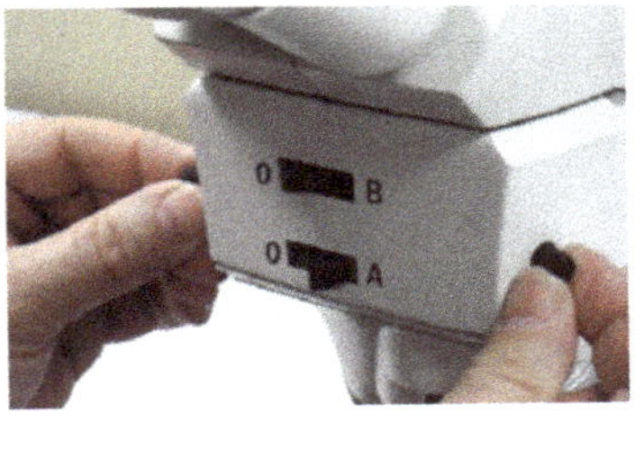

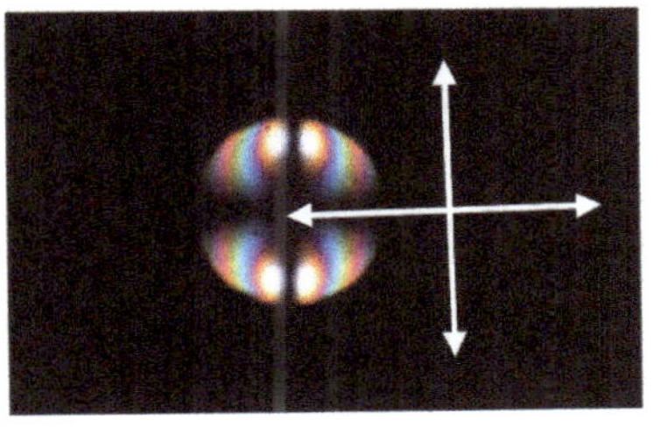

4) If you have an Analyzer/Bertrand Lens with a focusable Bertrand Lens, you can now turn the knob on the upper rocker to focus the Bertrand Lens image.

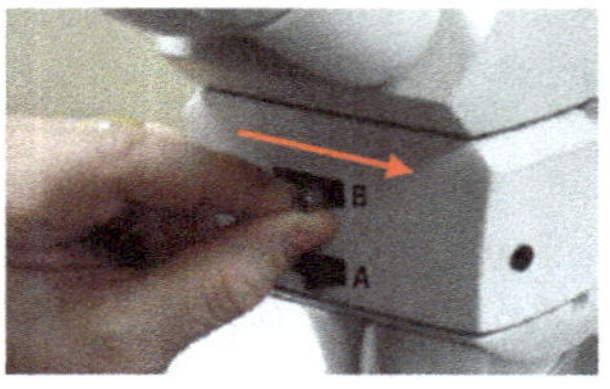

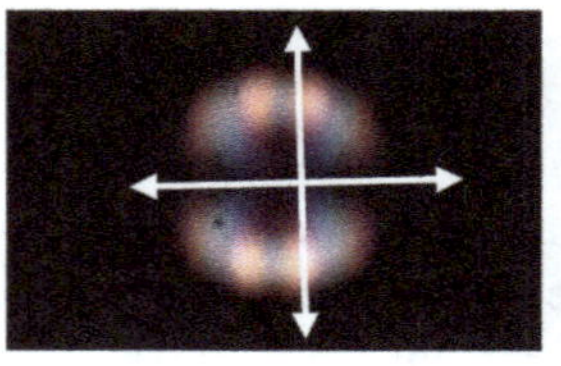

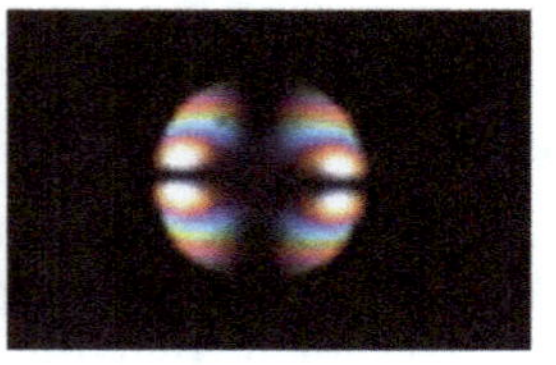

2.8 Reporting

1) Take images of a polymeric fiber and measure the diameter at different magnifications. Use the 'Process' module in LAS to measure simple dimensions. You need to repeat the measurement at least five times to get the average value and the standard deviation.
2) You need to include the images and the average diameter with standard deviation.
3) Discuss what is the best way to obtain most accurate measurement using an optical microscope.

Reference

[1] Murphy D 2001 *Fundamentals of Light Microscopy and Electronic Imaging* (New York: Wiley)

IOP Publishing

Through a Modern Microscope
Optical experiments on soft matter
Mohan Srinivasarao, Jung Ok Park and Karthik Nayani

Chapter 3

Experiment 2: Comparison of a real space image and q-space (diffraction) image for size measurement

3.1 Motivation

Let us look at figure 3.1. You must now be familiar with all the conjugate field planes and aperture planes. As you have seen in the previous labs, looking through the microscope with the Bertrand lens helps you see the conjugate aperture planes and hence the back focal plane. This chapter will cover the basics of the diffraction pattern formed in the back focal plane of a microscope and in particular for the case when we have diffraction gratings as the object. We will do so first with some basic background on diffraction. Most of the discussion is based on the bocks by Murphy and by Hecht [1–3]. Some parts of both references have been paraphrased for the sake of thoroughness.

3.2 Background

Francesco Grimaldi in the 1600s first published a detailed study on the deviation of light from rectilinear propagation, something which he called ‘diffracto’. The effects, however, are general characteristics of wave phenomena whenever a part of the wavefront is obstructed. The segments of the wavefront which propagate beyond the obstacle interfere and form what is the diffraction pattern. There is no fundamental distinction between diffraction and interference; however, it is somewhat customary to speak of interference when we consider a superposition of only a few waves and diffraction when we consider a large number of waves.

Although the quantum electrodynamics can explain diffraction of light thoroughly, and is well beyond the scope of the book, it is much more practical to understand diffraction using the Huygens–Fresnel principle despite the limitations of the principle. *The* ***Huygens–Fresnel principle*** *states that every unobstructed point of*

doi:10.1088/978-0-7503-1431-2ch3

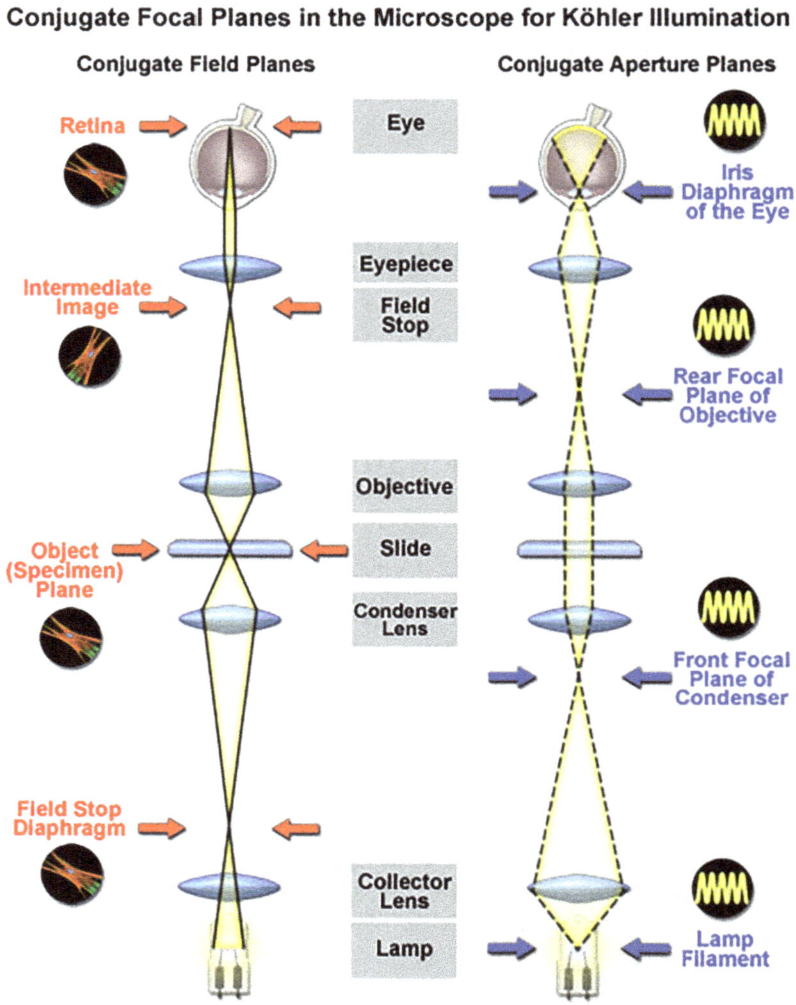

Figure 3.1. Illustrates the light paths in an optical microscope that is set up for Köhler illumination and the conjugate focal planes. Image credit: MolecularExpressions.com at Florida State University.

a wavefront, at a given instant, serves as a source of spherical secondary wavelets (with the same frequency as that of the primary wave). The amplitude of the optical field at any point beyond is the superposition of all these wavelets (considering their amplitudes and relative phases). Diffraction may then be envisioned as arising from the interaction of electromagnetic waves with some sort of physical obstruction. So, it is useful to think of Huygens–Fresnel principle as an outcome of the interaction of electromagnetic waves with matter.

Consider a screen placed in front of a light source. The electron cloud of each atom of the screen is set into vibration by the electric field of the incident radiation. The amplitude and phase of a particular oscillator within the screen are determined by the local electric field surrounding it. This in turn is a superposition of the incident field and the fields of all the other vibrating electrons. A large opaque screen with no aperture, be it made of black paper or aluminum foil, has one obvious effect: there is no optical field in the region beyond it. Electrons near the illuminated surface are driven into oscillation by the impinging light. They emit radiant energy, which is ultimately 'reflected' backward, absorbed by the material, or both. In any case, the incident wave and the electron-oscillator fields superimpose in such a way as to yield zero light at any point beyond the screen. This might seem a remarkably special balance, but it actually is not. If the incident wave were not canceled completely, it would propagate deeper into the material of the screen, exciting more electrons to radiate. This in turn would further weaken the wave until it ultimately vanished (if the screen were thick enough). Even an ordinarily opaque material such as silver, in the form of a sufficiently thin sheet, is partially transparent (recall the half-silvered mirror).

Now, remove a small disk-shaped segment from the center of the screen, so that light streams through the aperture. The oscillators that uniformly cover the disk are removed along with it, so the remaining electrons within the screen are no longer affected by them. As a first and certainly approximate approach, assume that the mutual interaction of the oscillators is essentially negligible, that is, the electrons in the screen are completely unaffected by the removal of the electrons in the disk. The field in the region beyond the aperture will then be that which existed before the removal of the disk, namely zero, minus the contribution from the disk alone. Except for the sign, it is as if the source and screen had been taken away, leaving only the oscillators on the disk, rather than vice versa. In other words, the diffraction field can be pictured as arising exclusively from a set of fictitious non-interacting oscillators distributed uniformly over the region of the aperture. This is the essence of the Huygens–Fresnel principle.

3.3 Experimental

With this basic understanding of diffraction and the principle of Huygens–Fresnel, we move on to diffraction from a diffraction grating.

A ***diffraction grating*** is a planar substrate containing numerous parallel linear grooves or rulings, light is strongly diffracted when the spacing between grooves is close to the wavelength of light (Why? Can you think about this based on your understanding of the Huygens–Fresnel principle?). If we illuminate the grating with a narrow beam from a monochromatic light source (again why monochromatic?) such as a laser pointer and project the diffracted light onto a screen, a bright, a central 0th-order spot is observed, flanked by one or more higher-order diffraction spots, the 1st-, 2nd-, 3rd-, etc-order diffraction maxima (see figure 3.2). The 0th-order spot is formed by the waves that pass straight through without diffraction. The diffraction spots identify unique path lengths which correspond to waves of same phases that interfere constructively. In the regions between the spots, the waves are out of phase and destructively interfere.

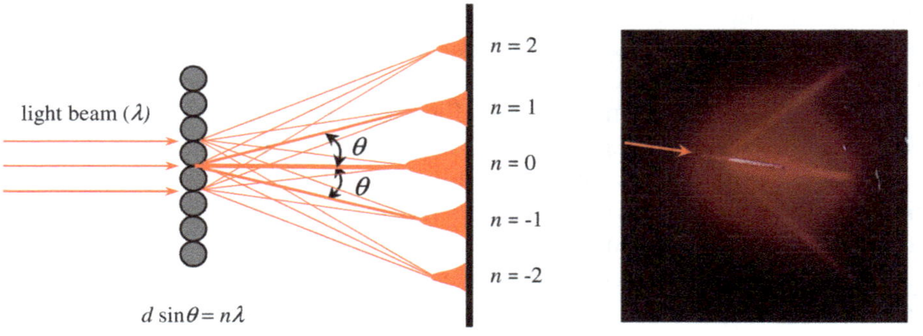

Figure 3.2. (Left) Intensity distribution of diffracted light from a diffraction grating; (right) a laser diffraction by a 1D grating, seen are the zero order and first order. The laser beam is indicated by the red arrow. Image credit: MolecularExpressions.com at Florida State University.

Under normal incidence, the diffraction angle θ of a grating is the angle subtended by the 0th- and 1st-order spots on the screen as seen from the grating. The right triangle containing θ at the screen is congruent with another triangle at the grating defined by the wavelength of illumination, λ, and the spacing between rulings in the grating, d. Thus, $\sin\theta = \lambda/d$, and reinforcement of diffraction spots occurs at locations having an integral number of wavelengths—that is, 1,2,3 and so on. Clearly we can calculate θ from the distance between the spots and the distance from the screen to the grating. With this, the spacing d of the rulings in the grating can be determined using **the grating equation,** $\boldsymbol{m\lambda = d \sin\theta}$, where m is the order of the diffraction.

Now with this knowledge, we can think of doing these following experiments with our microscopes:

1) Optimize the optical train by Köhler illumination and take a real space image of the diffraction grating using a suitable objective. Save an annotated image with a scale bar.
2) The grating equation above is based on the assumption of a normal incidence, and hence, it is important to have normal incidence of the beam on the microscope. What modification do you need to make to set up your microscope for normal incidence? Remember, the microscope is set for Köhler illumination!
3) Study the diffraction patterns formed by gratings on the back focal plane. Because we have white light source in the microscopes, what might the diffraction pattern look like? How would you expect the diffraction pattern to change as we change objectives with different magnification? (Think how numerical aperture can be related to diffraction. Remember in all our schematics and description, we assume the light is normally incident, hence the incoming light is *parallel*.) It is important to realize that the distance from the center of the optic axis of the microscope to the diffracted spot in the back focal plane provides an angle of the diffraction.
4) Measure the spacing of the gratings from the diffraction patterns.

5) Study diffraction patterns again using monochromatic light this time, by using interference filters. There are a number of interference filters that can be used to provide monochromatic light. Which color would you expect to have the largest diffraction angle?
6) If possible, measure the intensity profile across the diffraction spots. Most of the modern microscopes have a video camera or some recording device attached to it, and hence, it should be easy enough to measure the intensity profile across the diffraction spots.

3.4 Reporting

1) The real space image(s) of the grating(s) used, and their spacing you measured.
2) The diffraction image(s) of the grating(s) used, and their spacing measured from the diffraction images. Compare these measurements with the real space measurements and discuss the possible sources of the discrepancy if there is any.
3) Observations you made with respect to the questions asked in item 3) in the experimental section above.
4) Discussion on the advantages and drawback of using white light as incoming light.
5) The intensity profile across the diffraction spots, if possible.

3.5 Further suggested experiments

Now that you have used a diffraction grating that has a set of linear lines, it is possible to reproduce the experiments dealing with the resolution of the microscope. Referring back to section ……it should be evident that tilting the diffraction grating will allow one to limit the beams being collected by the objective—you could then ask yourself, is the grating being resolved under the conditions of tilting of the grating. One could also try to repeat the experiments described by Peter Evennett, that was discussed in part I of the text.

References

[1] Murphy D 2001Fundamentals of Light Microscopy and Electronic Imaging (New York: Wiley)
[2] Hecht E 1997Optics (Reading, MA: Addison-Wesley)
[3] http://micro.magnet.fsu.edu/primer/java/diffraction/basicdiffraction/

IOP Publishing

Chapter 4

Experiment 3: Microporous polymer films and 2D diffraction

4.1 Motivation

In many introductory materials science courses crystals, crystal structure and the determination of crystal structure are deemed important and are taught the x-ray crystallography as being the primary technique to determine crystal structure. Clearly, x-ray crystallography is quite useful in structure determination of large biomolecular systems such a proteins, DNA and RNA. Despite the importance, undergraduate students rarely have a chance to use the technique in a laboratory setting due to lack of facilities. And on top of it, the mathematics necessary for converting an x-ray diffraction pattern into a three-dimensional structure is not simple enough and hence, diminishes the understanding of the impact of x-ray crystallography experiments. The experiments presented below represent the ability of the students to discover single-crystal crystallography by preparing a two-dimensional array of holes that are ordered and studying how these structures diffract light and visualize the structure both in real space and in q-space.

In the previous experiment, experiment 2, we looked at a one-dimensional diffraction pattern from a diffraction grating. In this experiment, we will go a step further and analyze two-dimensional features by both real space imaging and diffraction patterns from well ordered pores in polymer films. Unlike previous experiment, in addition to analyzing samples, you will be preparing samples for the experiment, but luckily for us the method to prepare these regular ordered pores in polymer films is a remarkably simple process! This is also a model system for colloid crystallization; therefore, we will also observe the packing of the pores and their defects, as well as making measurements of the pore size by real space measurements coupled with q-space measurements.

4.2 Background

Breath figures, as the name suggests, refers to patterns of water drops one can see when you breathe over a cold object. The process is very similar to what happens

doi:10.1088/978-0-7503-1431-2ch4

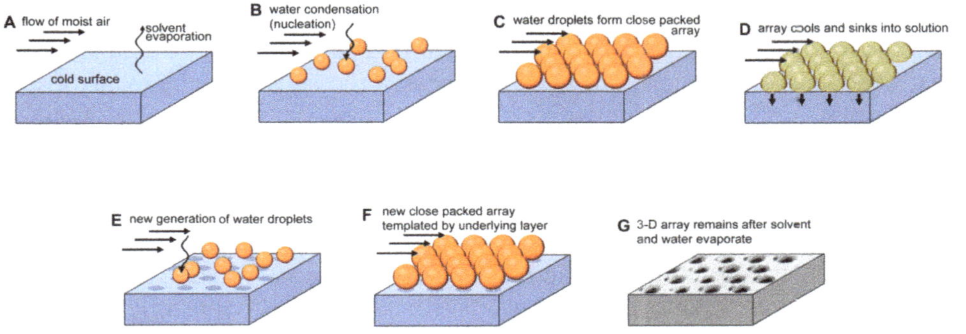

Figure 4.1. Mechanism for breath-figure-templated assembly suggested by Srinivasarao. Reproduced with permission from [1].

during dew condensation. However, what is really interesting is that, when you do breathe on evaporating dilute polymer solutions, you get water drops that are condensed from the mixture of the warm air and the solvent flux. These water drops grow and closely pack without any coalescence above the solution, and eventually sink into the solution. After the solvent and water drops are evaporated, a polymer film is left behind with hexagonally ordered close-packed pores typically in micron sizes.

Figure 4.1 [1] is a schematic description of how regularly spaced pores form over polymer films. As you can see from figure 4.1, the packing of the pores can be either two-dimensional or three-dimensional! The process can be applied to a wide variety of polymers, which opens potential applications like picoliter beakers, micolenses and photovoltaics, among others.

We will not discuss the mechanism of the formation of breath figures in this chapter, as it is beyond the scope of this chapter. However, a useful thing to think about is 'When you sprinkle water over a hot stove, notice how the water droplets never coalesce'. The mechanism in breath figure formation is very similar to what happens in the case of water droplets over a hot stove. (Think why that happens, you will better understand the mechanism of the breath figure process.)

4.3 Experimental

Typically you would prepare 1 wt % solution of mono carboxy terminated poly (styrene) (PS) in carbon disulphide (CS_2) for the experiments. We will use the PS with the molecular weight of 50 000 and/or 100 000 g mol^{-1}. Spread the drop of the solution on a clean microscope slide and blow moist air over the solution, perhaps as uniform as one is able to blow over the solution. In principle, the solution on a glass slide can just be left exposed to local air currents in the lab or on the microscope stage. Carbon disulphide is a volatile solvent which evaporates quickly and leaves behind a dried polymer film with ideally hexagonally packed monodispersed pores.

While the process for making these ordered arrays is fairly simple, the process is far from simple. On evaporation of the solvent, the solution cools due to *evaporative cooling*, which initiates *nucleation and growth* of water droplets, which *pack to form an ordered array*. We believe that micro-structuring via breath figure template has at least five steps that it goes through. The basic process is thought to involve the following stages: (i) Evaporative cooling of the polymer solution prompts the condensation of water vapor near the solution interface. (ii) The condensed water forms droplets, which grow in the presence of excess water vapor. (iii) Whilst enacting hard sphere-like interactions, the droplets pack to form a hexagonally ordered array of water droplets. (iv) On evaporation of most of the solvent, the water droplets become embedded in the polymer matrix. (v) Finally, evaporation of the residual solvent and the water droplets from the polymer substrate leads to the formation of an ordered array of porous film.

Figure 4.2 [2] shows the real space image and diffracted image from a breath figure sample (the settings, however, were more controlled than what we will be doing). One can see a beautiful six fold symmetry in the diffraction pattern, indicative of the hexagonal order of the breath figure assembly.

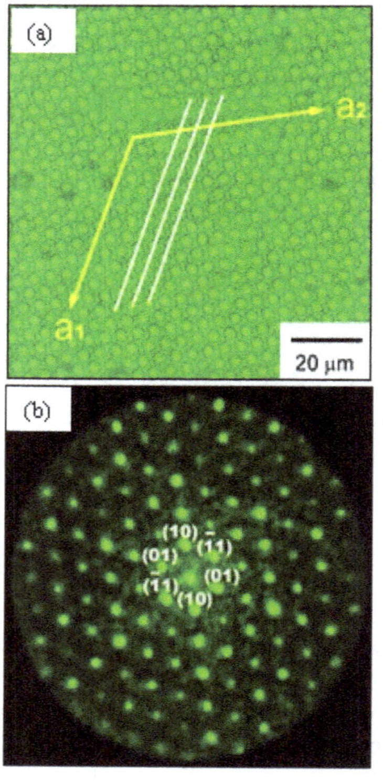

Figure 4.2. Direct-space image (a) and its diffraction pattern (b) of a polystyrene film, cast from a carbon disulfide solution (~1% by weight). Imaged by LEICA DMRX microscope with 40× dry objective (NA = 0.75) and an interference filter ($\lambda = 532$ nm). Miller indices (*hk*) of the first ring of diffraction spots are shown in (b) with the unit vector direction given in (a). The parallel white lines passing through the center of the air bubbles in (a) are the virtual lattice lines analogous to the parallel planes in a crystal. Reproduced from [2] with permission from the Royal Society of Chemistry.

In a similar vein to the calculations you performed to determine the spacing of the diffraction gratings, one can use the Bragg's law to determine the spacing of the holes in this regular array of holes. First, you need to measure the big circle radius, r_o, which is the maximum range in which your objective can capture the light, and the distances of each diffraction spot from the zeroth diffraction spot, r_{hk}. From the relation among r_o, θ_{NA} and r_{hk}, you can calculate d_{hk} using Bragg's law:

$$m\lambda = nd_{hk} \sin \theta_{hk}$$

The equation is similar to the 1D equation you used last week, but the d_{hk} term now represents the lattice spacing. For a layer of closely packed spheres, d_{hk} is the spacing between a set of parallel lines through the centers of the spheres where (hk) is the two-index specification of a set of parallel lines similar to the Miller indices which you might have been introduced to. So a (10) notation means that the normal to the plane is the x-axis which mean the reference is to the yz-plane and so on. One can determine the pore center-to-center distance, D, from d_{hk}, using the equation:

$$d_{hk} = \frac{D}{2}\left[\frac{3}{h^2 + hk + k^2}\right]^{1/2}$$

You will, however, notice that the sample you prepare will not be as ordered as the one in figure 4.2, because the conditions are not well controlled. That is not a bad thing, because it gives us an oppurtunity to study the defects in the crystal packing, in addition to determining d_{hk} and D. Try to study the defects in the system and see if you can find something peculiar about these defects.

4.4 Questions to answer and report

1) *Real space image analysis:*
   - Objective you used: Magnification/NA—also explain why NA is important in this case.
   - Average pore size (diameter) and size distribution.
   - Center-to-center distance.
   - Types of defects and change of packing characteristics due to defects.
   - Identify grain boundaries and as many types of defects possible (vacancy, edge dislocations).
2) *Diffraction pattern analysis*: This is done by viewing the microstructured polymer film in q-space with the use of a Bertrand lens in the optical path of the microscope.
   - Nature of the packing.
   - The maximum angle θ_{NA} that your objective can capture and the corresponding radius, in the back focal plane.
3) Average center-to-center spacing, D: You need to show your calculation by filling the table below.

h, k	r_{hk} (μm)	θ_{hk} (μm)	d_{hk} (μm)	D (μm)

- Repeat the experiments with white light, and perhaps in monochromatic light using a blue, green and red interference filter placed below the substage condenser. Repeat the calculations for the table for all three wavelengths.

4.4.1 Suggestions for further experiments

As should be obvious, the structure formation occurs to the polymer solution being cooled by evaporation, thus making the dew formation (in the broad sense a gas-to-liquid phase transition!) possible on a polymer substrate. However, one can easily perform an experiment where a solid substrate is cooled to a given temperature below room temperature, thus allowing the formation of what has traditionally been referred to as 'breath figures' analogous to the fogging of your glasses when you walk into a cold room from a humid outside. This experimental suggestion will allow you to monitor the nucleation and growth of water droplets on a cold surface. One could monitor the radius of the droplet as a function of time, as a function of the coverage of the surface and monitor what happens when the surface coverage exceeds about ~55%. We point you to a paper by Knobler and Beysens [3], for further study. Such studies have been used to understand the mechanism of chemical vapor deposition. Hence, while we have provided a very simple experiment, these experiments are profoundly useful in understanding a variety of complicated processes.

4.4.2 Historical analogue to the experiment

The structure formation that one will observe in the process of making these two-dimenisional structures are analogous to 'Bubble Rafts' that Sir William Lawrence Bragg and John Nye made in the late 1940s to demonstate crystal structure and the defects that form in atomic crystals. A bubble raft consists of an array of soap bubbles on the surface of water, often with the help of soaps. Bragg and Nye demonstrated that the bubbles act like atoms, diffusing, slipping and otherwise deforming in a way that likely models the deformation of close-packed crystals. These array also exhibit defects in an effort to pack efficiently. The ideal (lowest

Figure 4.3. An image of a bubble raft (bubble size ~1.5 mm) showing vacancies and an edge dislocation in the bottom right corner. This Bubblerraft2 image https://en.wikipedia.org/wiki/Bubble_raft has been obtained by the authors from the Wikimedia website where it was made available by Pieter Kuiper under a CC BY-SA 3.0 licence. It is included within this book on that basis. It is attributed to VladTarasov.

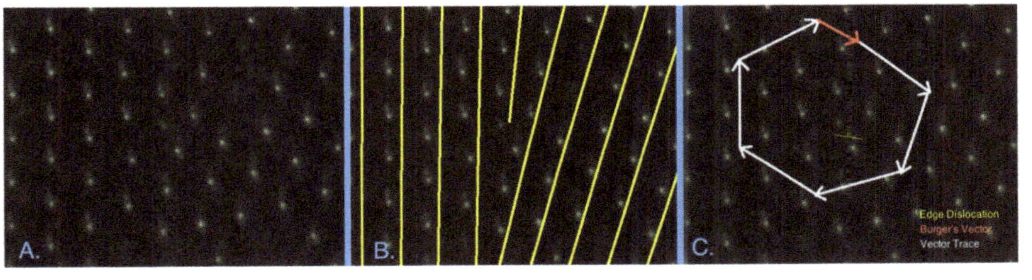

Figure 4.4. An image of a bubble raft (bubble size ~1.5 mm) containing an edge dislocation. This Dislocated Bubble image https://en.wikipedia.org/wiki/Bubble_raft has been obtained by the authors from the Wikimedia website where it was made available by Alex.penlington under a CC BY 3.0 licence. It is included within this book on that basis. It is attributed to Alex.penlington.

energy) state of the assembly would be perfectly regular single crystals, but just as in metals, the bubbles often form defects, grain boundaries and grains of crystalline order. Bubble rafts often display numerous phenomena seen in crystal lattices, that include such things as point defects (vacancies, substitutional impurities, and interstitial atoms, edge dislocations and grains. Some of these are illustrated in figures 4.3 and 4.4 dealing with bubble rafts.

4.4.3 Additional Experiments using colloids

Monodispersed spherical particles have been studied extensively for their ability to form two-dimensional crystals that pack with hexagonal symmetry. Since latex spheres with a high degree of monodispersity are available in a variety of sizes, they lend themselves to the formation and study of two-dimensional crystals. Such two-dimensional crystals can easily be formed by spreading a drop of latex spheres suspended in water on a glass slide, and then allowing for the evaporation to form a dry, two-dimensional crystal. It should be noted that the evaporation coupled with the capillary attraction drive the crystallization process, and should be easy to visualize on a light microscope.

To make such two-dimensional crystals, it is necessary to place a drop of latex sphere solutions on a glass slide and then spread the drop using a glass rod. The experiments can be done with water solutions of spheres in various sizes (ranging from submicron to 10 or so microns). Such solutions are available from a variety of sources. As the solvent evaporates, the spheres pack forming crystalline regions while observing this process in real-time on a microscope. Once the crystals are formed, mostly two-dimensional, one can observe the various defects, the diffraction pattern among other things as described previously for the formation of breath figure templated arrays.

References

[1] Srinivasarao M, Collings D, Philips A and Patel S 2001 *Science* **292** 79–83

[2] Song L, Sharma V, Park J O and Srinivasarao M 2010 *Soft Matter* **7** 1890–6

[3] Knobler C M and Beysens D 1988 Growth of breath figures on fluid surfaces *Europhys. Lett.* **6** 707–12

Beysens D and Knobler C M Growth of breath figures *Phys. Rev. Lett.* **57** 1433

IOP Publishing

Through a Modern Microscope
Optical experiments on soft matter
Mohan Srinivasarao, Jung Ok Park and Karthik Nayani

Chapter 5

Experiment 4: Liquid crystals, phase transitions, morphologies, and defects

5.1 Liquid crystal molecules and their phase transitions

Liquid crystals (LCs) are anisotropic fluids composed of rod-like or disk-like molecules. As the name implies, LCs exhibit some characteristics that are common to both normal isotropic liquids and solid crystals. It is precisely this ambivalence that leads to a great diversity of phenomena related to surface effects. Such surface effects have been studied to a large extent using nematic LCs. The first reason for this is the simplicity of their structure, where the local order aligning the molecules parallel to each other can extend to macroscopic dimensions, primarily driven by the surface. The second, and perhaps the most important, reason is that nematic LCs are primarily used in display devices, where the surface properties of the cell play an essential role in the working of a display device.

In a nematic phase, the molecules exhibit long-range orientational order but no positional order, except perhaps near a surface which perturbs the structure of the nematic phase. This orientational order is characterized by the special direction of local order, known as the director, $\boldsymbol{n}$, and the order parameter, $\boldsymbol{S}$, which indicates the degree to which the molecules are aligned about the director. However, the direction of $\boldsymbol{n}$ is not unique in the absence of proper surface treatment or without the application of an external field. Because the nematic phase possesses orientational order and is uniaxial in nature, it is often characterized by high birefringence. *Hence, polarized light microscopy is an essential tool in probing and understanding LCs.*

Therefore, this laboratory is to observe the morphologies/textures and topological defects of nematic LCs and phase transitions of LCs as a function of temperature. Topological defects are stable configurations of matter formed upon phase transitions. The study of topological defects provides a unique link to the

doi:10.1088/978-0-7503-1431-2ch5

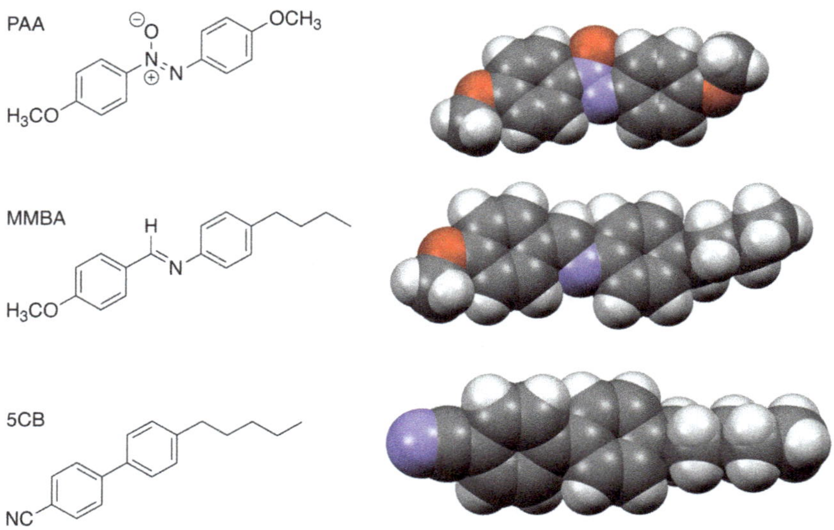

Figure 5.1. Typical compounds forming nematic mesophases. (top) (PAA) p-azoxyanisole. From a rough steric point of view, this is a rigid rod of length ~ 20 Å and width ~ 5 Å. The nematic state is found at high temperatures (between 116 °C and 135 °C at atmospheric pressure). (middle) (MMBA) *N*-(*p*-methoxybenzylidene)-*p*-butylaniline. The nematic state is found at room temperatures (between 20 °C and 47 °C). This molecule does not have good chemical stability. (bottom) (5CB) 4-pentyl-4′-cyanobiphenyl. The nematic state is found at room temperatures (between 24 °C and 35 °C). Redrawn after [2] Image credit: Denis Andrienko.

physics of the very early Universe, for those who are keen to pursue this topic further. Nematic LCs exhibit an array of topological defects. We will observe the different kinds of defect structures LC often display. The reader is encouraged to refer to the classic text *The Physics of Liquid Crystals*, by P G de Gennes [1].

Many compounds are known to form a nematic mesophase. A few typical molecules displaying LC phases are sketched in figure 5.1. From a steric point of view, molecules are rigid rods with the breadth-to-width ratio from 3:1 to 20:1 [2].

When these compounds are heated from the solid (=crystal) state, they undergo phase transitions at different temperatures, and the number of different liquid crystalline phases are dependent on the molecular structure. A few examples are shown in figure 5.2 with their phase transition temperatures. The temperature for nematic–isotropic transition is called T_{NI}, the *nematic–isotropic transition temperature*.

In order to be used in LCD industry, any LC or LC mixture has to maintain its nematic phase in a wide temperature range. E7 (Merck) is one of those mixtures, whose composition is shown in table 5.1 [3] and its property as a display material is shown in table 5.2 [3].

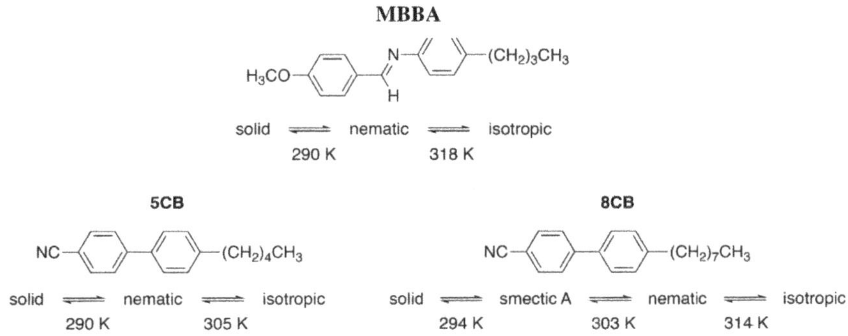

Figure 5.2. MBBA, 5CB, and 8CB each possesses two benzene rings each and a long hydrocarbon tail. The =N– linkage in MBBA is largely responsible for its negative dielectric anisotropy. 5CB and 8CB each possesses a highly polar CN group. The number before the 'CB' (cyanobiphenyl) indicates the number of CH groups in the hydrocarbon tail. LC phases are indicated along with their respective transition temperatures. Redrawn after [3].

Table 5.1. Composition of E7 [3].

Chemical compound (% in E7)		Name	Individual nematic range (°C)
51%	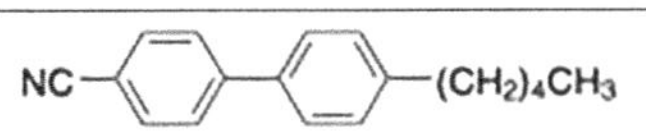	5CB	24 → 35.3
25%	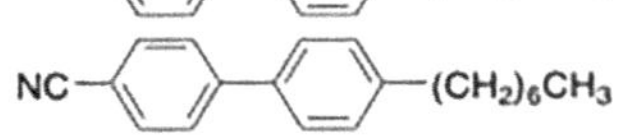	7CB	30 → 42.8
16%	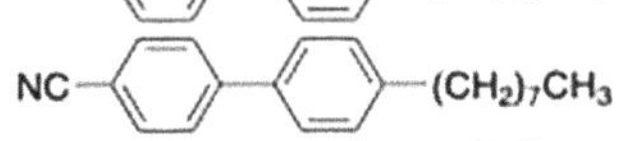	M24	67 → 80
8%	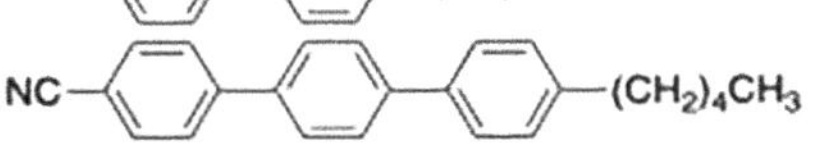	T15	131 → 240

Table 5.2. E7 cyanobiphenyls as a display material [3].

Property	Requirement	E7
Nematic range	−10 °C to +60 °C	−9 °C to +59 °C
Threshold voltage	<1.0 V	1.2 V
Resistivity	$>5 \times 10^{10}$ ohm cm	5×10^{10} ohm cm
Response time	<50 ms	35 ms
Color	None	None

5.2 Anisotropic optical properties of LC

The LC phases are anisotropic in nature and many of their properties also display anisotropy: their molecular shape, optical property, dielectric property, magnetic susceptibility, etc. This lab mainly involves the optical anisotropy. While some of the following has been discussed in Part I of the text, we repeat some aspects of the discussion below, to help with the flow as well as this part being used as a self-contained description of the experiments.

As discussed in earlier sections on polarized light microscopy, when light travels in an anisotropic medium, it splits into two rays with different velocities of propagation whose polarization directions are orthogonal to each other. One of the rays does not change its direction, hence it is called '*the ordinary ray*', and its velocity is associated with the **ordinary** refractive index, n_o. The other ray changes its direction, hence called '*the **extraordinary** ray*', and its velocity is associated with the extraordinary refractive index, n_e as schematically illustrated in figure 5.3). The distance of separation between the ordinary and extraordinary ray increases as the thickness of the uniaxial crystal or material increases.

Figure 5.4 illustrates a birefringent crystal placed between two crossed polarizers at 45° to the plane of polarizer, and is indicated by the arrows next to the polarizer and analyzer labels.

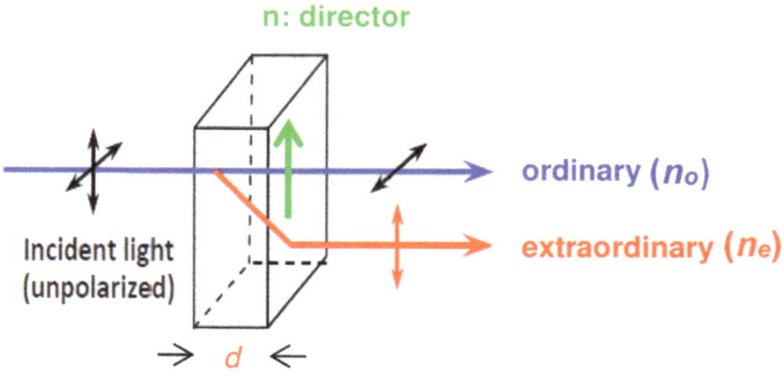

Figure 5.3. A diagram showing that the unpolarized incoming light rays are separated and displaced with orthogonal polarizations when they pass through a birefringent material.

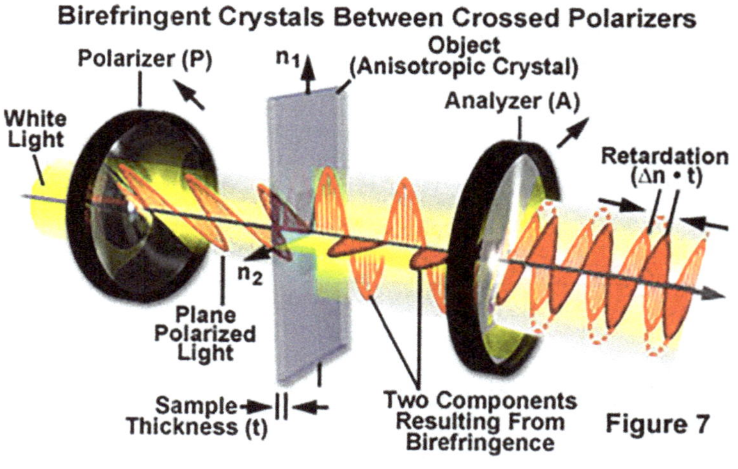

Figure 5.4. A diagram to show the retardation between the ordinary ray and the extraordinary ray generated by an optically anisotropic sample and the crossed polarizers. Reproduced with permission from 'MolecularExpressions.com at Florida State University'.

When white light enters the polarizer in figure 5.4, it is polarized in the direction indicated by the arrow and is arbitrarily represented by a 'red' sinusoidal light wave. This polarized light entering the anisotropic crystal is split into two separate components vibrating parallel to the crystallographic axes and perpendicular to each other (arbitrarily, the 'blue' and 'red' light waves). The polarized light waves then pass through the analyzer (whose polarization position is indicated by the arrow, which passes only those components of the light waves that are parallel to the polarization direction of the analyzer.

Let's consider the phase relationship and velocity differences between the ordinary and extraordinary rays after they pass through a birefringent material. Because of the difference in refractive indices, one ray will pass through the crystal more slowly than the other ray. Therefore, the velocity of the slower ray will be **retarded** with respect to the faster ray. This *retardation* can be quantified by:

$$\textbf{Retardation } (\Gamma) = \textbf{thickness } (t) \times \textbf{birefringence } (B)$$

$$\Gamma = t|n_e - n_o|$$

Because one wave is retarded with respect to the other, interference (either constructive or destructive) occurs between the waves as they pass through the analyzer. The retardation has the unit of the length and its magnitude depends on the sample thickness and birefringence. The retardation defines the interference color of thin LC films under crossed polarizers when the LC film is illuminated with white light. Since the birefringence, Δn, changes according to the molecular alignment changes due to the fluid flow and/or temperature gradient, etc, even when the film thickness remains the same, the color of a thin LC film varies locally within the sample. This is why the LCs under crossed polarizers show a variety of colors that constantly changes until the equilibrium is reached. The Michel–Levy chart (see figure 5.5) summarizes the color change due to the change in the

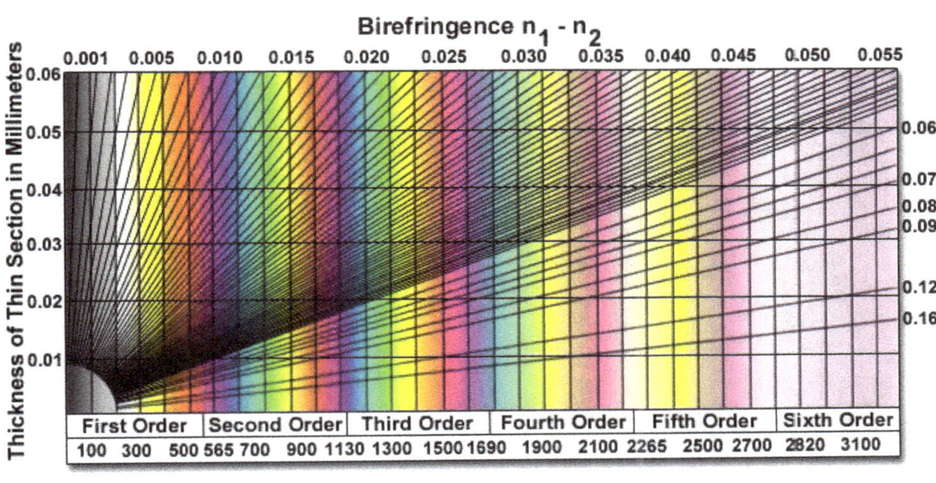

Figure 5.5. Michel–Levy chart showing the interference color changes of optically anisotropic materials upon the change in the sample thickness and the birefringence. The chart is from Zeiss.

birefringence and the sample thickness. This is a simpler discussion than that provided in the earlier section dealing the colors of the Michel–Levy chart.

5.3 Liquid crystal morphology: interference colors, characteristic textures, and defects

5.3.1 Nematic LC

Typical optical textures under the crossed polarizers of thin nematic LC films are shown in figure 5.6.

The Schlieren texture of a nematic film is due to the topological defects across which the LC molecules are differently orientated as shown in figure 5.7.

5.3.2 Cholesteric LC

Nematic LCs doped with enantiomorphic (= chiral) materials become cholesterics of long (but finite) pitch. The molecules forming this phase are always optically active, i.e. they have distinct right- and left-handed forms. The pitch of the common cholesterics is of the order of several hundred nanometers, and thus comparable with

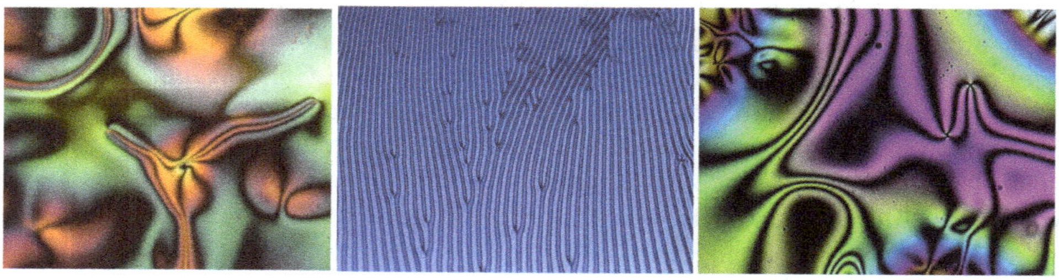

Figure 5.6. (Left) *Schlieren* texture of a nematic film with surface point defects (boojums). (Middle) Thin nematic film on isotropic surface (on the surface of glycerol): one-dimensional periodicity. (Right) Schlieren texture showing two point defects.

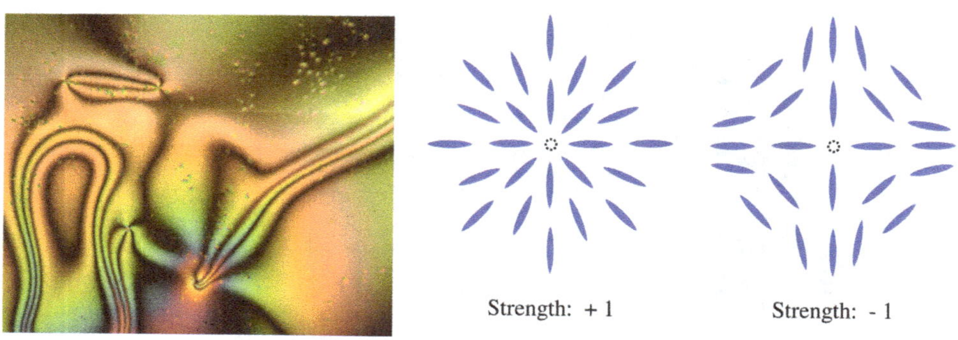

Figure 5.7. Texture from topological defects of LC using polarized optical microscope. Clearly seen are point defects and they usually come as pairs, as a +1 and a −1, as seen in the image.

the wavelength of visible light. The spiral arrangement is responsible for the characteristic colors of cholesterics in reflection (through Bragg reflection by the periodic structure) and their very large rotatory power. The pitch can be quite sensitive to temperature, flow, chemical composition, and applied magnetic or electric field. Typical cholesteric textures are shown in figure 5.8.

5.3.3 Smectic LC

In general, when placed between glass slides, the smectic layers, preserving their thickness, become distorted and can slide over one another in order to adjust to the surface conditions. The optical properties (focal conic texture) of the smectic state arise from these distortions of the layers. Typical textures formed by smectics are shown in figure 5.9.

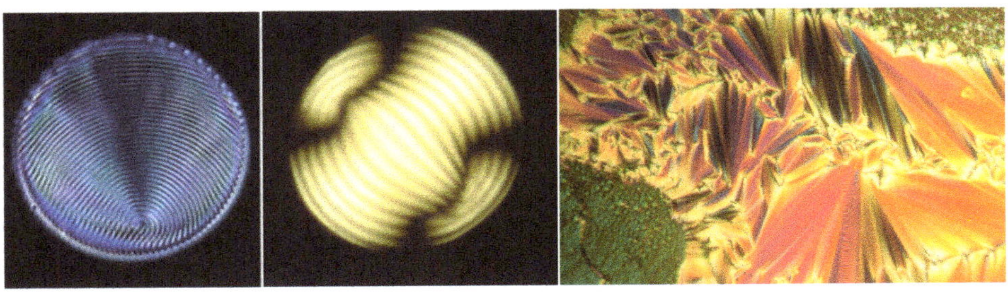

Figure 5.8. (Left) Cholesteric fingerprint texture confined to a spherical droplet composed of a nematic fluid with a chiral additive. The line pattern is due to the helical structure of the cholesteric phase, the spacing between the lines provide a measure of half the pitch. (Middle) A cholesteric droplet formed by chitin nanocrystals suspended in water and confined to a sphere. A short-pitch cholesteric LC in Grandjean texture or standing helix texture, viewed between crossed polarizers.

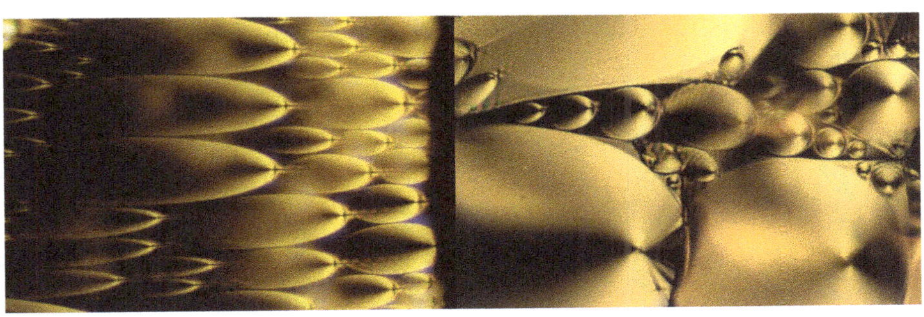

Figure 5.9. Focal conic structures formed by a typical smectic A LC. Images courtesy of Oleg Lavrentovich, at Kent State University (http://www.lavrentovichgroup.com/textures.html).

5.3.4 Discotic LC

Disk-shaped mesogens can orient themselves in a layer-like fashion known as the discotic nematic phase. If the disks pack into stacks, the phase is called a discotic columnar. The columns themselves may be organized into rectangular or hexagonal arrays (see figures 5.10–5.12).

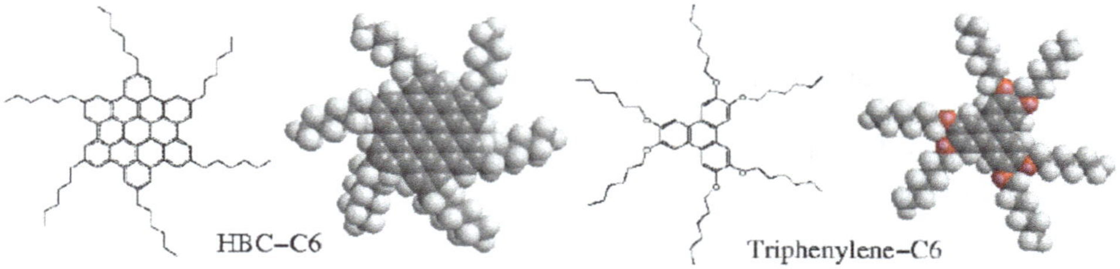

Figure 5.10. Typical discotics: derivative of a hexabenzocoronene and 2,3,6,7,10,11-hexakishexyl-oxytriphenylene. K(70 K) → Col_h(100 K) → I. From [2] (http://creativecommons.org/licenses/by/4.0/). Image credit: Denis Andrienko.

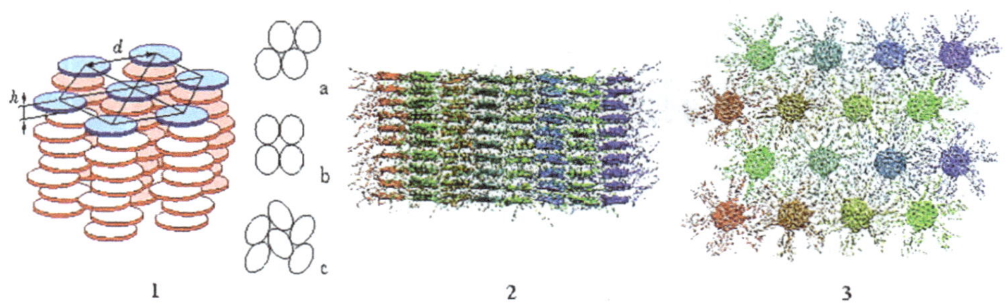

Figure 5.11. (1) Columnar phase formed by the disc-shaped molecules and the most common arrangements of columns in two-dimensional lattices: (a) hexagonal, (b) rectangular, and (c) herringbone. (2, 3) MD simulation results: snapshot of the hexabenzocoronene system with the C12 side chains. Aromatic cores are highlighted. From [2] (http://creativecommons.org/licenses/by/4.0/). Image credit: Denis Andrienko.

Figure 5.12. (Left) Columnar phase of lyotropic LC formed by DSCG in water. (Middle) Texture of a columnar phase formed by lyotropic LC formed by Sunset Yellow in water. (Right) A biphasic texture comprising of an isotropic phase + a columnar phase of Sunset Yellow in water. Images courtesy of Oleg Lavrentovich, at Kent State University (http://www.lavrentovichgroup.com/textures.html).

5.4 Experiments

Thus far we have provided a simplistic view of LCs and how a nematic fluid interacts with an incident polarized light. The following experiments are meant to familiarize you with how to see the defects in the system as well as the structures observed during a phase transition between the nematic–isotropic phases. We will also observe the effect of the surfaces on the LC morphology and defects, and the phase transition under various surface conditions using crossed polarizers on a light microscope.

5.4.1 Questions to ask

Does the sample extinguish light under crossed polarizers? How many times does it extinguish light as a function of rotating the sample (zero to 360°)? Take images to show these.

1) Use a twisted nematic (TN) cell:
    a) Fill E7 nematic LC in a TN cell by capillary suction.
    b) Heat your sample on a hot plate to reach an isotropic phase → observe it while it cools down to a nematic phase under crossed polarizers → record a few images as a function of time during cooling.
    c) After the LC alignment reaches the final equilibrium state, answer the questions above. Are there any defects? Take a few images to show these.
2) Use a *planar nematic LC sample*:
    a) Place a tiny drop of E7 LC near an opening of a planar cell and wait until the LC fills the cell.
    b) Repeat the steps shown in (1a–c) above.
3) Use an *open cell LC sample*:
    a) Take a glass slide and rub it with a sheet of Kim-wipes or a velvet cloth only in a given direction. Rub the glass plates a sufficient number of times (maybe 10) but only in one direction (not back and forth). Place a tiny drop of LC on a rubbed glass slide and gently spread it as thin as possible. Do not use any cover slide. Observe and document the defects that you find—and try to interpret what those defects mean.
    b) Repeat the steps shown in (1a–c) above. Make as many observations as possible and attempt to interpret what you see based on what has been discussed thus far.
4) Use an *LC sample on glycerol surface*:
    a) Use a very thin layer of glycerol on a non-rubbed glass slide and then deposit a tiny amount of nematic LC on glycerol.
    b) Repeat the steps shown in (1a–c) above.
5) Use an *CLC sample with glycerol surface*:
    a) Use a very thin layer of CLC on a non-rubbed glass slide. Take images of the as-prepared sample and measure the pitch.
    b) Heat it to isotropic phase, cool it to room temperature, take images and measure the pitch. Discuss any change in pitch.

5.4.2 Suggestions for further experiments

Now that one has looked at a variety of defects during heating and cooling through a phase transition, one can do a few more experiments to further understand how nematics behave under confinement. Take a capillary tube with a diameter of about 50 μm or so, and treat the capillary with a surfactant solution. The surfactant usually promotes alignment of the optic axis of the system perpendicular to the surface of the capillary. Once, this has been done, one can fill the capillary with various mixtures of nematic fluids that one can buy and make observations under a microscope with crossed polarizers. Should the director be normal everywhere, observation under crossed polarizers ought to be dark with a line defect running through the center of the cylinder; however, a line defect along the entire length of the capillary is expensive energetically, and hence, the director escapes along the length of the capillary, in what is referred to as 'escape in the third dimension'. Make observations of the appearance along the length of the capillary and try to understand what the nematic is doing. These were some of the early experiments that were conducted on nematic fluids to understand their behavior. This has an added advantage in that you will traveling down the path of pioneers in LCs, a worthwhile pursuit.

Another experiment would be to create a monodomain, meaning the entire nematic sample is pointing in a given direction. One can assemble cells with two glass plates that have been rubbed in a unidirectional fashion and fill them with a nematic fluid. This can be filled with a variety of LCs, and one can do conoscopic experiments as described in the section in Part I dealing with conoscopy and optical crystallography. Using the monodomains, one can try to measure the birefringence (D_n) and the two refractive indicies, n_o and n_e, quantitatively. One can do this as a function of temperature as well, thus providing a measure of the order parameter of the system.

Another experiment to attempt would be to create a wedge cell with a both planar and homeotropic alignments. This cell can be used to interrogate the double refraction of nematic fluids.

References

[1] de Gennes P G and Prost J 1993 The Physics of Liquid Crystals (Oxford University Press)

[2] Andrienko D 2018 Introduction to liquid crystals *J. Mol. Liq.* **267** 520–41

[3] Kawamoto H 2002 The history of liquid-crystal displays *Proc. IEEE* **90** 460

IOP Publishing

Through a Modern Microscope

Optical experiments on soft matter

Mohan Srinivasarao, Jung Ok Park and Karthik Nayani

Chapter 6

Experiment 5: Features in liquid crystal display (LCD)

6.1 Background

6.1.1 What is liquid crystal (LC)?

In addition to the three well-known forms of matter, i.e., solid, liquid and gas, certain materials show intermediate phases between crystalline solid and isotropic liquid, which are *thermodynamically stable*. They possess many of the mechanical properties of liquid such as high fluidity, formation and coalescence of droplets. At the same time, they are similar to crystals in that they exhibit anisotropy in their optical, mechanical, electrical and magnetic properties. Therefore, they are termed as ***liquid crystal (LC)*** phases [1–4].

Crystal ←→ Liquid crystal ←→ Isotropic

(Cr) (mesophase) (I)

Temperature increase →

The molecules in a liquid crystal phase are often shaped like rods or plates or some other anisotropic shape that encourage them to align collectively along a certain direction, often denoted as the **director** (see figure 6.1). The characteristic feature of LCs is the presence of long-range *orientational order* in the arrangement of constituent molecules, and sometimes one- or two-dimensional quasi-long-range *translational* or *positional order*.

doi:10.1088/978-0-7503-1431-2ch6

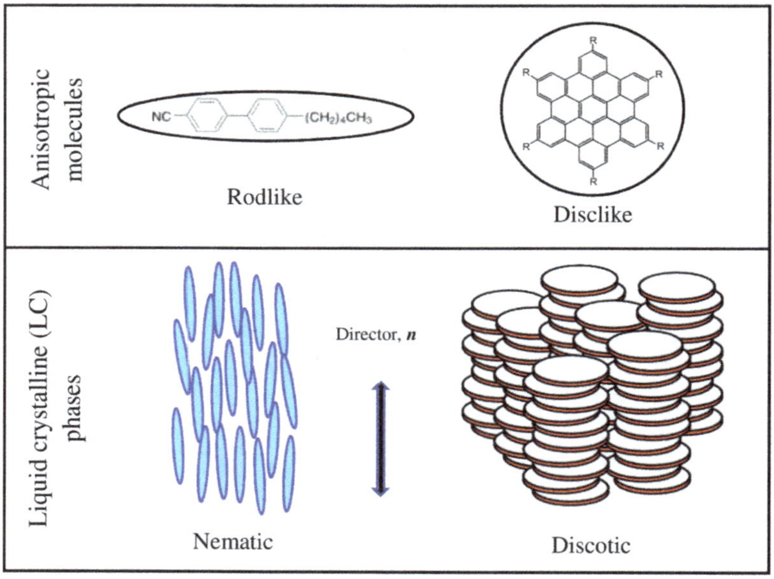

Figure 6.1. Schematic illustration of the self-assembly of a rod-like molecule (left) and a disc-like molecule forming a nematic phase (images below the molecules). The director, ***n*** is parallel to the long axis of the rod-like molecules while it is normal to the plane of the disc-like molecules.

6.1.2 Classification of liquid crystals

Liquid crystals are divided into different groups, based on how they form LC phases:

- **Thermotropic LC phases** are formed in a certain temperature range hence the prefix *thermo-*, referring to phase transitions caused by a temperature change.
- **Lyotropic LC phases** form in a solution, and thus concentration controls the liquid crystallinity (hence, prefix *lyo-*, referring to concentration) in addition to temperature. Thermotropic LCs do not need solvent to form an LC phase.

In our experiments, we are going to deal with a rod-like thermotropic liquid crystal.

There are a number of LC phases, based on how the molecules are aligned, but we will learn about three particular types of liquid crystalline phases: nematic, smectic and chiral nematic (cholesteric) phase (see figure 6.2).

- **Nematic phase** is the simplest phase consisting of molecules with long-range one-dimensional orientational order: meaning, all the molecules are pointing along a particular direction on average. This direction is denoted as a *director*, ***n***.
- In **smectic phases**, which are sometimes found at lower temperatures than the nematic, molecules form well-defined layers that can slide over one another. The smectics are thus positionally ordered in one direction. In a given layer, the molecules are aligned as in a nematic phase.
- Chiral molecules cause a twist in the nematic structures. Such a helical phase is called ***chiral nematic*** (or ***cholesteric***) LC. Locally, the cholesterics are very

Figure 6.2. Three characteristic phases of LCs with different structural orders: Nematic (left); smectic A and C (middle); and chiral nematic (right). These LiquidCrystal images https://en.wikipedia.org/wiki/Liquid_crystal have been obtained by the authors from the Wikimedia website where they are made available by Kebes under a CC BY-SA 3.0 licence. It is included within this book on that basis. The are attributed to Kebes.

similar to the nematic LCs. They consist of quasi-nematic layers, whose individual directors are turned by a fixed angle on proceeding from one layer to the next. The layers turned by an angle of 2π are equivalent; the distance between these two is defined as the *pitch* of the helical structures.

- In a thermotropic system, order is changed as a function of temperature so that phase transitions occur with increasing temperature. At high temperatures, the thermal motion will destroy the ordering of the liquid crystal phase, thus forming an isotropic phase. The transition temperature between the nematic phase and the isotropic phase is called *nematic–isotropic transition temperature*, T_{NI}.

6.1.3 Nature of *crystallinity* (*anisotropic* physical properties)

1) **Optical anisotropy**: LC is a birefringent medium, as already discussed, where the refractive index is direction dependent. Light *polarized parallel to the director* (= extraordinary ray) 'sees' one refractive index, which is called *extraordinary refractive index* (n_e). Light polarized *perpendicular to the director* (= ordinary ray) 'sees' a different refractive index, called *ordinary refractive index* (n_o), as calcite does. The difference in these refractive indecies is called the **birefringence**, $\Delta n = n_e - n_o$. Some examples of the birefringent materials are shown in table 6.1.

   Let's consider the phase relationship and velocity differences between the ordinary and extraordinary rays after they pass through a birefringent sample. Because of the difference in refractive indices, one ray will pass through the sample at a slower rate than the other ray, hence the velocity of the slower ray will be **retarded** with respect to the faster ray. This retardation can be quantified using the following equation:

$$\textbf{Retardation } (\boldsymbol{\Gamma}) = \textbf{thickness } (\boldsymbol{t}) \times \textbf{birefringence } (|\boldsymbol{\Delta n}|)$$

$$\boldsymbol{\Gamma} = \boldsymbol{t}\ |\boldsymbol{n}_e - \boldsymbol{n}_o|$$

Table 6.1. Refractive indicies and the birefringence of common birefringent materials [data from https://en.wikipedia.org/wiki/Birefringence].

Crystal	n_o	n_e	Δn
Calcite	1.6584	1.4864	−0.1720
Quartz	1.5443	1.5534	0.0091
Sodium nitrate	1.5854	1.3369	−0.2485
Ice	1.309	1.313	0.0014
Rutile (TiO_2)	2.616	2.903	0.287

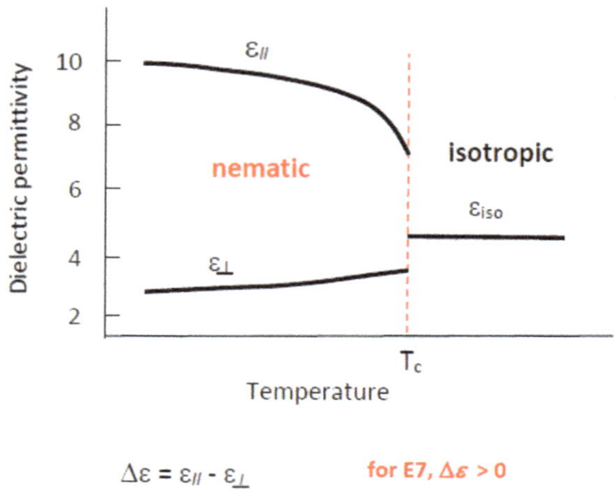

Figure 6.3. Schematic depiction of the dielectric constant (both along and perpendicular to the director) as a function of temperature for a typical nematic fluid in both the nematic and isotropic phases. Courtesy: Oleg Lavrentovich, Kent State University.

Chiral nematic and certain smectic phases (like smectic C*) are *optically active*, i.e., a *linearly polarized light is rotated* as it passes through such a medium.

2) **Dielectric anisotropy**: Dielectric properties of LCs are related to the response of LC molecules to an applied electric field. The dielectric susceptibility (ε) is a physical quantity that describes how molecules in a dielectric medium polarize in response to an applied electric field. LCs have dielectric anisotropy, which means the dielectric susceptibility is different in parallel ($\varepsilon_{\|}$) and perpendicular ($\varepsilon_{\hat{a}}$) direction to the director. Therefore, the difference, $\Delta\varepsilon = (\varepsilon_{\|/} - \varepsilon_{\hat{a}}$), determines the molecules' reorientation (parallel or perpendicular to the applied field) (see figure 6.3) on the application of an electric field.

6.1.4 Principle of liquid crystal display

LCD is a flat panel display that uses the light modulating properties of LCs. LCs do not emit light themselves, but simply act as an on/off switch. LCD is an electrically modulated optical device made up of any number of segments filled with LCs and arrayed in front of a light source (backlight). The colors are generated using red–green–blue (RGB) color filters and modulation of the light intensity passing through these filters. Each filter has its own thin-film transistor (TFT) that modulates the light intensity.

1) **Operating principle** (example is a twisted nematic (TN) mode)
   - ✓ Liquid crystals in a nematic phase align along the grooves on the substrate (in this case, glass, which is usually coated with polymeric, usually a polyimidie alignment layer). When coming into contact with a grooved surface in a fixed direction, **liquid crystal molecules align parallel to the grooves**, thus minimizing the free energy of the system.
   - ✓ By arranging the grooves of the two substrates (upper and lower ones) at 90° to each other, molecules continuously twist up to 90° between two substrates. **When no voltage is applied**, polarized light travels through this molecular arrangement. The light also gets 'twisted' or 'the polarization is rotated' as it passes through the twisted LCs. The polarization is rotated by 90 degrees as it follows the twist of the molecules. Polarized light eventually passes through the analyzer, if the LC thickness (t) satisfies the condition (= adiabatic limit) of $\boldsymbol{t\Delta n > \lambda/2}$, where $t\Delta n$ is retardation as described above. See figure 6.4 (left).

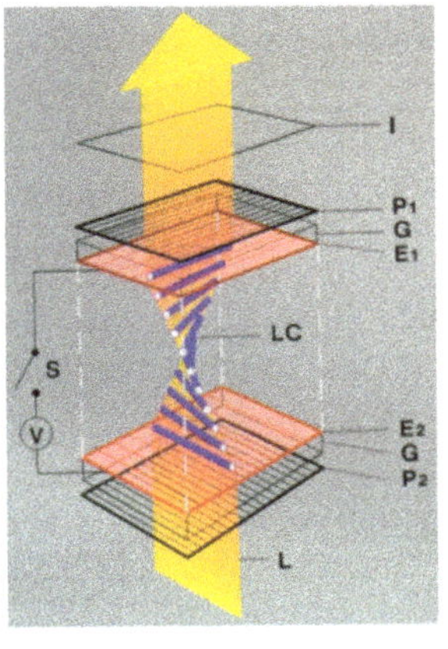

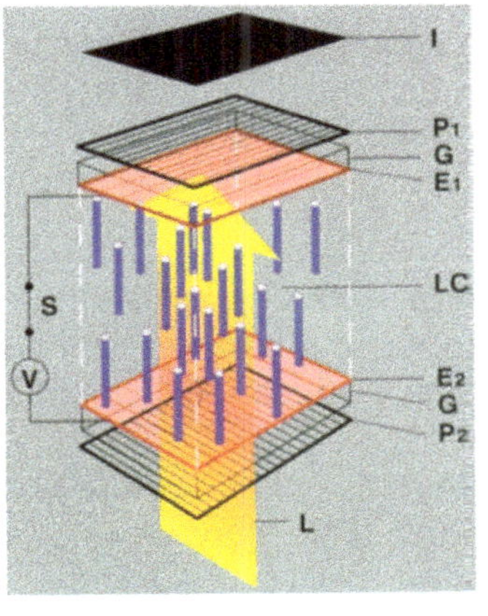

Figure 6.4. On a twisted nematic configuration, (left) when no voltage is applied, the polarized light gets twisted; (right) when voltage is applied, molecules align along with the electric field, and no light passes through. These TN-LCD-schematic-MS-208kB-OFF/ON images have been obtained by the authors from the Wikimedia website where they are made available by Em3rgent0rdr under a CC BY-SA 3.0 licence. They are included within this book on that basis. They are attributed to M. Schadt.

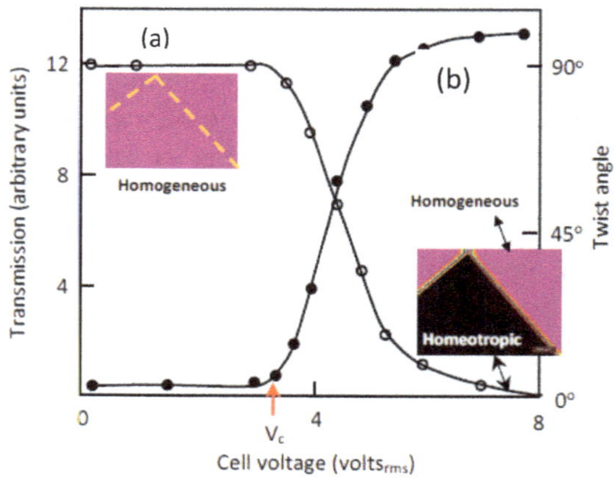

Figure 6.5. (a) Light transmission and rotation angle of linearly polarized light versus voltage for the cyano-Schiff's base mixture at room temperature (at 1 kHz) based on the data extracted from [5]. Also displayed are microscopic images showing a single pixel as it is changed from homogenous at low voltage to homeotropic at high voltage, viewed under crossed polarizers.

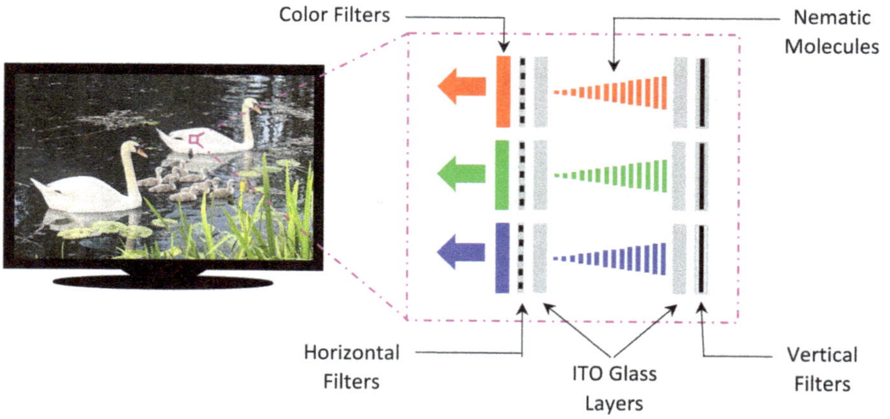

Figure 6.6. TFT LCD to express full color using RGB color filters, TN LC on/off switch and TFT to control the light intensity for each pixel.

✓ **When voltage is applied**, molecules align along with the electric field, and possess no twist. There is no polarization rotation, and no light passes through. See figure 6.4 (right).

Figure 6.5 illustrates the transmission and the twist angle as a function of the applied field, which is the basis of the on–off function of the LCD.

6.1.5 Full color LCD

So far, we have seen that the LC molecules in LCD work as on–off switches or intensity controllers in combination with TFTs, depending on the applied electric field strength. The colors of the full color LCD are achieved by means of RGB color filters that are individually coated on each TFT element, as shown in figure 6.6.

6.2 Experiments

1) We will discuss why LC appears milky-white, and observe the phase transition to isotropic phase (clearing), upon heating, with our naked eye. Upon cooling, the LC returns to turbid state.
2) Pixel per inch (PPI) and resolution of cell phones: Find an area of white color in any image on your cell phone, take a picture of the color filters, and calculate the PPI and the resolution of the LCD of your cell phone. Compare it with those of your lab mates' phones and the corresponding cell phone maker's advertised values. Report all these values.
3) We will see how LCD of our cell phone operates to make different images. Download the color wheel in figure 6.7 [6], and find out how the color mixings are made using RGB scheme to express these colors. Capture images of your LCD pixels and quantify the RGB intensities of these pixels for these color wheels. Report the results.
4) Using a TN LC cell, take images of your cell as a function of applied voltage strength and measure the transmitted light intensity. Make a plot of the intensity versus the field strength.

Figure 6.7. An example of color wheels [6]. https://en.wikipedia.org/wiki/Color_wheel# ~:text=The typical artists' paint or,colors of light being absorbed.

References

[1] Chandrasekhar S 1992 *Liquid Crystals* 2nd edn (Cambridge: Cambridge University Press)

[2] Collings P J and Hird M 1997 *Introduction to Liquid Crystals: Chemistry and Physics* (London: Taylor and Francis)

[3] de Gennes P G and Prost J 1993 *The Physics of Liquid Crystals* 2nd edn (Oxford: Oxford University Press)

[4] Kawamoto H 2002 The history of liquid-crystal displays *Proc. IEEE* **90** 460

[5] Schadt M and Helfrich W 1971 Voltage-dependent optical activity of a twisted nematic liquid crystal *Appl. Phys. Lett.* **18** 127–8

[6] https://en.wikipedia.org/wiki/Color_wheel#:~:text=The typical artists' paint or,colors of light being absorbed

Chapter 7

Experiment 6: Polymer crystallization

7.1 Introduction

Many of the materials that are in use today, in the sense of the word, are substances that possess physical properties that are of interest. Glass, for example, has optical properties that makes it desirable for a variety uses that include data communication by means of optical fibers. Polymers constitute another class of materials, along with glasses (amorphous materials), metals, ceramics and semi-conductors. Many classes of materials form solids with crystalline order. In general, most crystalline substances, under some conditions, when crystallizing, often grow with a spherical form that includes disparate substances as simple salts, organic molecules, metals, metal alloys, polymers, and proteins. While the spherical form is only an idealization, however, on observation with the help of a light microscope, such growing crystals are remarkably spherical, and are described as spherulites. Often such observations are made by confining the materials between glass substrates, and thus, spherulites grow as radial disks with cylindrical symmetry. It must be evident that such a form can exist only from successive branching of the nucleus that forms when a liquid of the material is cooled below its melting point. *In this experiment, we will utilize common polymers to observe the process of a liquid becoming a crystalline solid.* In other words, the material, a polymer, undergoes a phase transition from being a liquid to a crystalline solid.

7.2 Background

Polymers are ubiquitous in our everyday life due to their broad spectrum of properties, be it synthetically made or naturally occurring. Common examples are synthetic polymers such as polystyrene, used in coffee cups, or polyethylene to natural biopolymers such as DNA and proteins that are fundamental to life as we know it. The word 'polymer' denotes a material that is a large molecule, or

doi:10.1088/978-0-7503-1431-2ch7

macromolecule consisting of many repeat units that are covalently linked. The origin of the term polymer is from Greek, simply implying 'many (*polus*)' parts (*meros*). To be precise, a polymer is a molecule consisting of repeat units linked by covalent bonds, and the best-known example is polyethylene, made from *ethylene* ($CH_2{=}CH_2$), called a *monomer*, to form a polymeric chain whose structure can be represented by $-[CH_2-CH_2]_n-$ where n represents the number of repeat units linked by covalent bonds, or the degree of polymerization, and can be large. That polymers are covalently linked macromolecular structures was proposed in 1920 by Hermann Staudinger, and such linear arrangements are often referred to as *polymer backbone*.

In the most ideal case, a polymer backbone can be considered to be a connected set of carbon atoms that are free to assume a large number of confirmations due to the inherent ability of a carbon–carbon bond to rotate in space. Hence, the shape of a single polymer chain can be altered simply by the rotation of a carbon–carbon bond. However, there are a large number of such bonds in a polymer chain, and each has the ability to rotate independently. Thus, a polymer chain samples a large number of confirmations and hence is often referred to as a random coil. For example, a polyethylene chain with a molecular weight of $280\,000\ \mathrm{g\,mol^{-1}}$ has a degree of polymerization of $n = 10^4$, thus providing a large number of confirmations for the chain to sample in three-dimensional space. Hence, it becomes necessary to use statistical methods to estimate the dimensions of such a chain, and is often modeled by a random walk or random flight model or the ideal polymer chain.

Despite the dramatic, and rather unrealistic approximations at first sight, there are a number of instances such as in polymer melts or polymer solutions in the vicinity of a special temperature (q-temperature where two-body interactions vanish) where the polymer chains can be successfully described by ideal chains—those are beyond the scope here; however, in the following we provide a glimpse of what an ideal chain is, as it does provide a reasonable estimate of the dimensions of the polymer chains in melt.

The model of a random flight chain or an ideal chain or a random walk chain, the monomers are of fixed step length, l, and each step's orientation or the direction is entirely independent of the direction and position of the next step. Hence, if the polymer backbone has N monomers, the total length of the polymer is simply given by

$$L = N{\cdot}l,$$

where L is referred to as the 'contour length', and l is the length of each step. Thus, when there are no interactions between monomers, the energy of the chain is independent of its shape, and hence, all of the confirmations are equally likely to occur as the polymer chain explores space. Let us then say $\vec{R}$ is the total end-to-end vector and $\vec{r}_1, \ldots, \vec{r}_N$, the vectors corresponding to the individual steps (the monomers in a chain), where N is large. The figure below illustrates the path of a random walk for a (short) ideal chain:

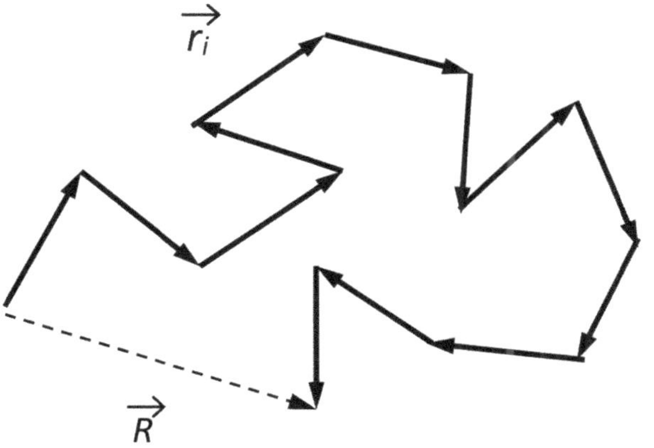

If one defines the end-to-end distance as the vector sum of all the individual vectors, then it is obvious because of the randomness, the average end-to-end distance is identically equal to zero, as shown below:

$$\langle \vec{R} \rangle = \left\langle \sum_{i=1}^{n} \vec{r_i} \right\rangle = 0$$

This is not a useful result for determining the dimensions of a chain. So, in search of the first nonzero value of the end-to-end distance, we arrive at the *mean square end-to-end distance*:

$$\left\langle \vec{R^2} \right\rangle = \left\langle \sum_{i=1}^{n}\sum_{j=1}^{n} \vec{r_i}\vec{r_j} \right\rangle = \mathrm{Nl}^2$$

Thus, the root-mean-square end-to-end distance of an ideal or random flight chain is simply as follows.

RMS end-to-end distance is $\sqrt{R^2} = \sqrt{N}l$. Hence, the average rms end-to-end distance, $\langle R_{\mathrm{rms}} \rangle$ is simply proportional to $\sqrt{N}$, thus showing that the size of the chain is much smaller than the contour length, which is proportional to N.

Thus, most of the polymers in the melt have long molecular chains in the form of irregular entangled coils, a random coil. Upon cooling, some polymers retain such a disordered structure and thus convert into amorphous solids (= amorphous polymer), while the chains in others polymers, rearrange and form partly ordered regions typically with a micrometer dimension (= (semi-)crystalline polymers) (see figure 7.1 [1]). Common examples of semi-crystalline polymers are listed in table 7.1 [1]. Such semi-crystalline polymers possess less than 100% crystallinity, with the density of the crystalline region being higher than the amorphous region due to the very compact and regular packing of the molecular chains in crystalline region, as illustrated in figure 7.1(b).

Although it is energetically favorable for the polymer chains to align parallel to each other, due to the chain entanglement, the polymer chains are both aligned and

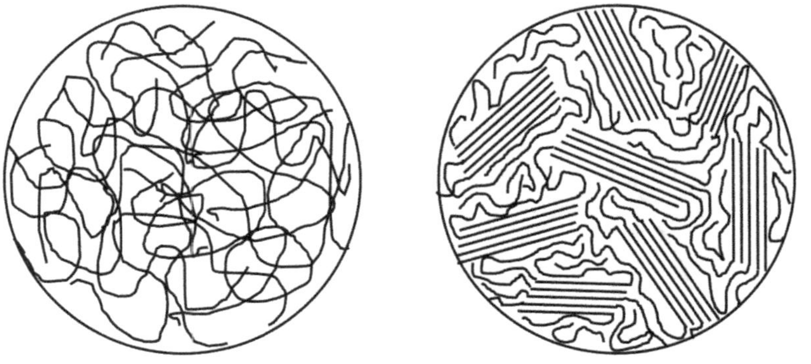

Figure 7.1. The arrangement of molecular chains in (a) amorphous and (b) (semi-) crystalline polymers (redrawn after [1]). The (semi-) crystalline polymers have both amorphous and crystallization regions.

Table 7.1. Degree of crystallinity (D), densities of crystalline (ρ_c) and amorphous ($\rho_{a,}$) polymers [1].

Polymer	D (%)	ρ_c (g cm^{-3})	ρ_a (g cm^{-3})
Nylon (PA66 and PA6)	35–45	1.24	1.08
Polyoxymethylene (POM)	70–80	1.54	1.28
Polyethylene terephthalate (PET)	30–40	1.50	1.33
Polybutylene terephthalate (PBT)	40–50	—	—
Polytetrafluoroethylene (PTFE)	60–80	2.35	2.00
Isotactic polypropylene (i-PP)	70–80	0.95	0.85
Atactic polypropylene (a-PP)	~0	—	—
High-density polyethylene (HDPE)	70–80	1.0	0.85
Low-density polyethylene (LDPE)	45–55	1.0	0.85

folded within the ordered regions, called lamellae, which compose larger spheroidal structures named spherulites (see figure 7.2).

Polymer crystallization can occur by various pathways: cooling from the melt, mechanical stretching or solvent evaporation. When crystallized rather slowly from a solvent, one gets single crystals that have been studied extensively, and will not be discussed here. Upon crystallization, polymers change their optical, mechanical, thermal and chemical properties.

Polymer crystallization also depends on their molecular structure and presence of straight chains with regularly spaced side groups that facilitate crystallization. For example, crystallization occurs more easily in isotactic polypropylene (i-PP) than in the atactic polypropylene (a-PP) form (see figure 7.3 [1]). Atactic polymers crystallize when the side groups are very small, as in polyvinyl, and don't crystallize in the case of large substituents like in rubber or silicones.

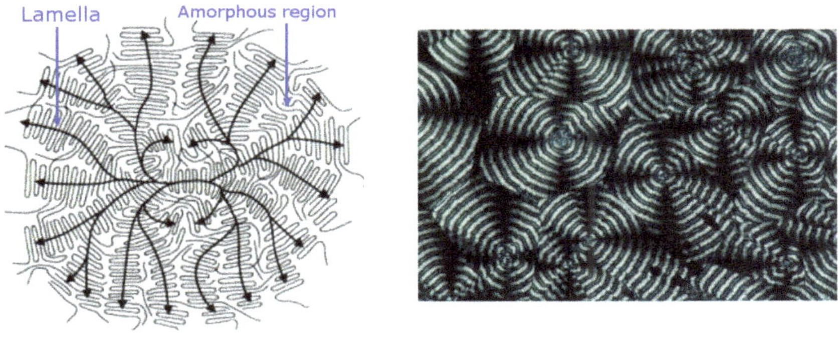

Figure 7.2. (Left) Schematic model of a spherulite of a polymer. Black arrows indicate direction of molecular alignment, and hence the polymer chains fold perpendicular to the lamella. This Spherulite2 images has been obtained by the authors from the Wikimedia website where it was made available by Materialscientist under a CC BY-SA 3.0 licence. It is included within this book on that basis. It is attributed to Materialscientist. (Right) Banded (or ringed) spherulites of poly-hydroxybutryate imaged in a light microscope under crossed polarizers. The image clearly displays a dark cross, often known a maltese cross (courtesy Professor Bart Kahr, NYU). https://en.wikipedia.org/wiki/Crystallization_of_polymers; Creative Commons Attribution-ShareAlike 4.0 license.

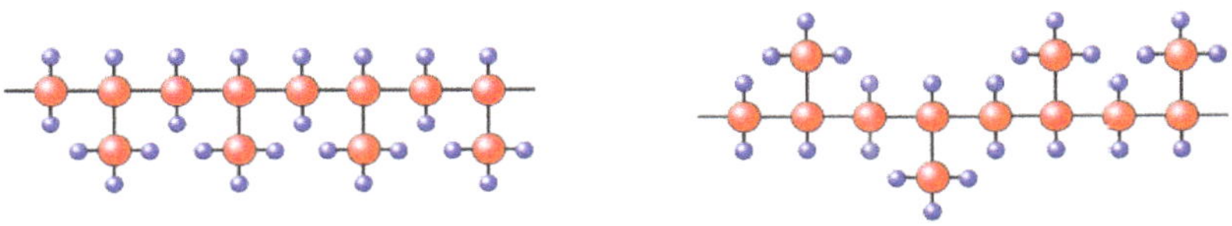

Figure 7.3. The structure of isotactic polypropylene (i-PP) and atactic polypropylene (a-PP) [1]. This Isotactic & Atactic polypropylene image https://en.wikipedia.org/wiki/Crystallization_of_polymers has been obtained by the authors from the Wikimedia website, where it is stated to have been released into the public domain. It is included within this book on that basis.

7.3 Motivation

There are a large number of materials that crystallize; however, when studying crystallization of small molecules, the process generally occurs too rapidly to study the kinetics of the process in detail. Therefore, the advantage of using polymers as model systems to study crystallization is that the relatively slow kinetics of polymer crystallization allows one to use an optical microscope and make observations of crystal growth, the growth rates as well as the morphology of the system. It also presents one with the opportunity to study the phase transition or phase transformations, where a liquid polymer melt undergoes a liquid–solid phase transition. It should be noted that such transformations are widely studied in materials science and engineering, and are also taught in most curricula.

7.4 Nucleation and growth

It is well known that the most accepted path for crystallization (or condensation from a gas to a liquid) is by a two-step process: nucleation and growth. The driving force for such a transition is that the free energy of liquid and that of the solid are equal at the melting temperature (T_m) of the solid. Below the melting temperature, the free energy of solid is lower than that of the liquid and hence, the liquid solidifies forming a crystalline solid. The free energies of the liquid and solid can be written as:

$$G_l = H_l - TS_l$$

$$G_s = H_s - TS_s$$

where H_l, H_s, S_l and S_s are the liquid and solid enthalpies, and entropies, respectively. In general $H_l > H_s$ and $S_l > S_s$ due to the fact that the molecules have a higher energy in the liquid state in comparison to the solid state. At the melting temperature $G_l = G_s$ and hence, one can write the *latent heat of solidification* L as

$$L = H_l - H_s = T_m \Delta S$$

where $\Delta S = S_l - S_s$, is the *entropy change on solidification*. Therefore, solidification occurs only below the T_m, i.e. at an *undercooling* $\Delta T = T_m - T$, with the *driving force for solidification* ΔG given by

$$\Delta G = L\Delta T / T_m$$

Thus, the driving force for solidification increases linearly with undercooling below the melting temperature.

The classical theory for crystal nucleation proposes that a small crystal nucleates in an undercooled liquid phase because of random fluctuations. The free energy of the system is maximized when a critical nucleus size is reached, and further growth corresponds to lowering of free energy, as illustrated in figure 7.4.

Nucleation occurs spontaneously through random fluctuations in the parent, liquid phase, through two pathways: *homogenous nucleation*, when a nucleus forms without any external nucleating sites, and *heterogenous nucleation*, which is aided by a defect or foreign substances.

Assuming homogenous nucleation, a small spherical (an assumption) crystal nucleates spontaneously with a radius r in the liquid phase. For this to occur, there are two components to the free energy of nucleation: the volume free energy that works towards the formation of a stable nucleus, and since on nucleation one is creating a surface, a surface free energy that works against the nucleation/growth of

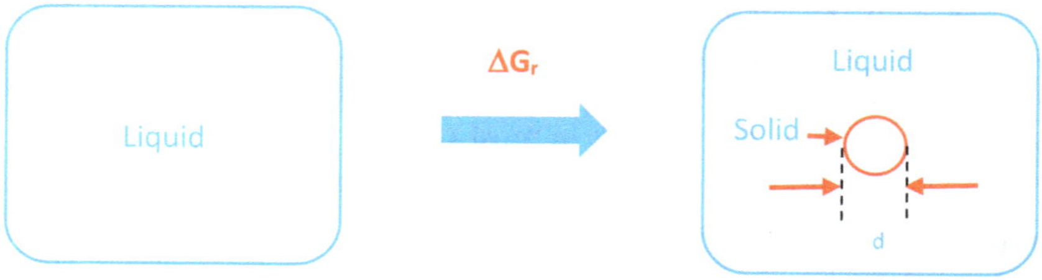

Figure 7.4. Schematic illustration of a liquid transforming to a solid via nucleation in the liquid.

the solid phase. Hence, the change in Gibbs free energy for this process can be written as

$$\Delta G = (4/3)\pi r^3 \Delta G_v + 4\pi r^2 \gamma = \text{Volume term} + \text{Surface term}$$

The volume free energy change ΔG_V, is always negative, while the interfacial energy γ is always positive. When the nuclei are small the surface term dominates (due to higher surface-to-volume ratio) and when the nuclei are large the volume term dominates, thus making the nuclei viable and able to continue to grow to form the new solid phase. The competition between the two terms is schematically illustrated in figure 7.5, where it is clear once a critical nucleus size (r^*) is reached, the solid phase continues to grow as the free energy change is favorable for growth. Hence, at some intermediate value of r, the free energy goes through a maximum, and the probability of formation of new nucleus goes through a minimum. Thus, there exists a least-probable nucleus size (r^*), the critical nucleus size, with the highest value of ΔG, where

$$r^* = \frac{d\Delta G}{dr} = 0 \Longrightarrow \frac{2\gamma}{\Delta G_V}$$

Addition of more molecules to the critical nucleus (size $= r^*$) deceases the free energy, and hence, the structure grows as a function of time.

When the free energy becomes negative, the stable nuclei then enter the second step of the phase transition: the growth phase.

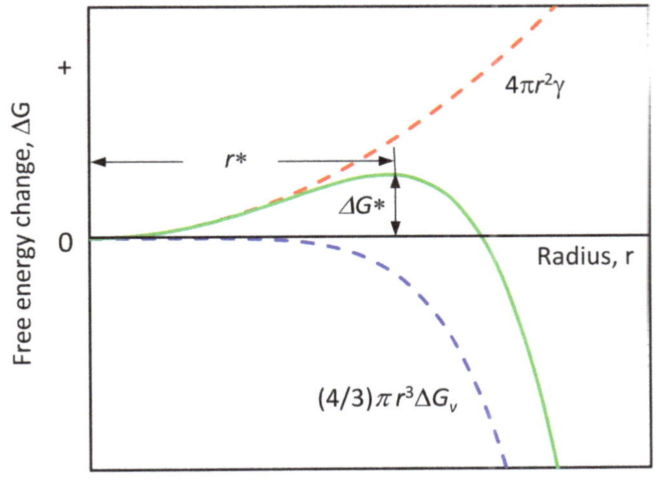

Figure 7.5. Schematic illustration of the free energy change for nucleation: the green curve represents the total Gibbs free energy as a function of the radius of the nucleus. The total free energy is the sum of the two components, the surface term (red dotted line, quadratic in r) and the bulk term (blue dotted line, scales as the cube of the radius, r) which scales with volume and is always negative, illustrating the origins of the energy barrier for nucleation.

The time dependence of such growth has often been described by what is referred to as the Avrami equation. This equation was originally developed to model the progression of phase transformations, but is now widely used for a variety of processes. It provides a functional form for the fraction of the material that is transformed as a function of time $\{f(t)\}$, and is a sigmoidal function and is written as

$$f(t) = 1 - \exp(kt^n)$$

where k, n are constants that are extracted from the model, and has its origins in the classical nucleation theory. Figure 7.6 illustrates a typical plot that such an equation provides. It should be noted that the above expression is also referred to as the Johnson–Mehl–Avrami–Kolmogorov (JMAK) equation, referring to the work done independently by Johnson and Mehl [2], Avrami [3–5] and Kolmogorov [6]. The interested reader can find more details elsewhere [7]. This curve illustrates that the initial (phase) transformation is slow, followed by an acceleration of the growth of the phase and eventually slowing down towards the completion of the transformation. In the case of nucleation and growth of spherulites from the melt, the nuclei form then grow, eventually impinging on other spherulites (which are solids) at which point most of the material might have been transformed into the solid. The question of what phase nucleates (i.e. the structure) is a much more complicated one, in particular for long molecular units like polymers, where the state corresponding to the lowest free energy is almost never attainable due to kinetic contributions during the growth phase [8].

From experiment 4 that dealt with liquid crystals, it is evident that any material possessing anisotropy in its optical properties (birefringence) can be analyzed and studied using a polarizing light microscope. In a melt, most of the linear flexible polymer chains assume random configurations, and hence, it is isotropic in its optical properties, and one expects no transmission of light through crossed polarizers. As we start cooling from the melt phase, the crystals nucleate and

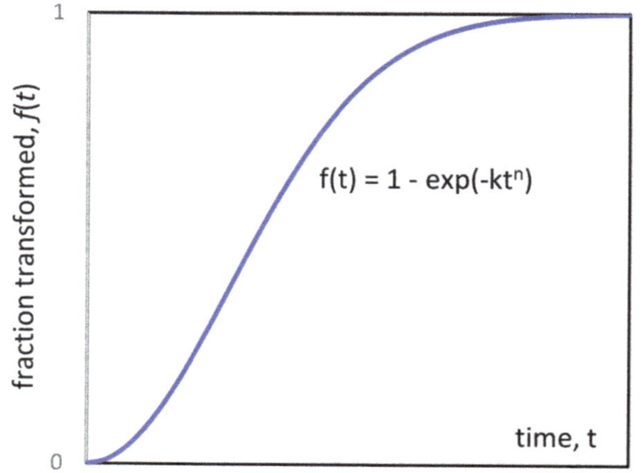

Figure 7.6. Schematic illustration of the progress of the phase transformation at constant temperature, as dictated by the Avrami equation.

grow, and, due to the anisotropy in the system, they become visible under crossed polarizes, thus allowing for a study of their growth process.

So as discussed above, the polymer crystallization can be thought of as limited by the diffusion of a polymer chain to the crystal interface and the formation of a critical nucleus size. A simple model capturing the growth rate is the Turnbull–Fisher equation [9], written as

$$G(T) = G_0 \mathrm{e}^{-E/kT}\, \mathrm{e}^{-\varphi/kT}$$

where $G(T)$ represents the growth rate of the spherulites. The first exponential term represents the activation energy barrier for transporting the polymer chain to the crystal melt interface and the second term is the activation energy required to form the critical sized nucleus. T is the temperature at which the crystallization is being carried out. So, we should expect a Gaussian-like growth function. However, the growth rates also usually have an exponential dependence with time [9].

There have been several attempts to model the kinetics of polymer crystallization. One of the most widely used models is the Avrami theory [3–5], that has been detailed above. This simple theory treats crystallization as a macroscopic phenomenon in which nuclei appear at random points within the polymer volume and give rise to crystals that grow at a certain rate around the nuclei [3–5].

Figure 7.7 shows the spherulites of poly(ethylene oxide) (PEO) observed under cross polarizers as a function of time. As can be seen from the images, the nucleation starts and the spherulites grow as a function time. One can also clearly see another generation of nuclei nucleating and impinging on the first generation but once they impinge, the clarity of the Maltese cross is lost. In the end, the field of view is completely occupied by the spherulitic structures.

The molecular organization in ultrathin polymer films (thicknesses less than 1000 Å) and thin polymer films (thicknesses between 1000 and 10 000 Å) may be

Figure 7.7. Polarized microscope images of spherulites of PEO (MW = 500 000), nucleated at 40 °C, showing the progress of crystallization as a function of crystallization duration at arbitrary intervals. The scale shown in the first of the images in 20 μm.

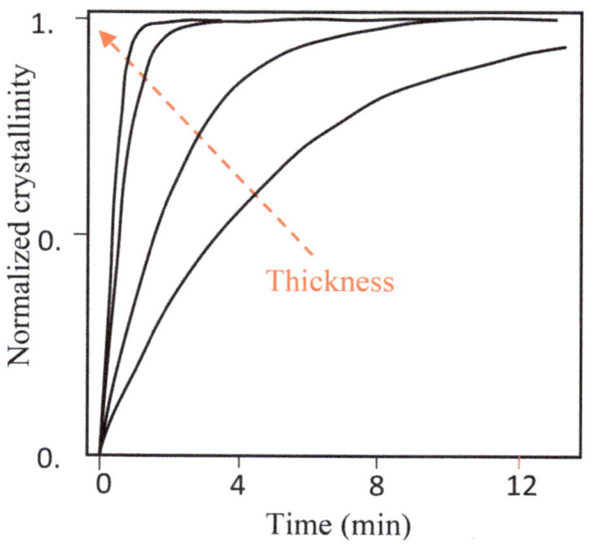

Figure 7.8. Schematic of the kinetics of crystallization at 0 °C of spin-cast poly(di-*n*-hexyl silane) (PD6S) films of thickness (increasing right to left) 95 Å, 160 Å, 300 Å, and 500 Å. The lines shown are traces of the nonlinear fits of the Avrami equation to the normalized experimental crystallinity data. Redrawn schematically without the experimental data points, after [10].

substantially different from that of bulk polymers, resulting in important differences in thermophysical properties. Figure 7.8 shows the results (schematically drawn) of a series of experiments at a crystallization temperature of 0 °C [7]. *The crystallization rate decreases dramatically as the film thickness decreases*, particularly for thicknesses <300 Å. Moreover, over the range from 15 °C to −5 °C, *the rate of crystallization increased as the crystallization temperature decreased.* The solid lines are the best fits of the Avrami equation to the experimental data. The fitting parameters for a film 500 Å thick vary with temperature, suggesting that the crystallization changes from a one-dimensional process to a three-dimensional process at a temperature between 5 °C and 0 °C, possibly due to a transition from bulk to heterogeneous nucleation as the temperature increases [10].

7.5 Experimental

The basic experiment is simple; we heat the polymer over its melting temperature and allow it to cool. A more controlled experiment would have involved cooling to a particular temperature and studying the growth isothermally.

However, we can explore crystallization for different cooling rates and see the growth rates and the spherulite structures that evolve. In order to achieve different cooling rates, we will heat the PEO (MW = 100 000 g mol^{-1}) at different temperatures, using three hot plates, and monitor the spherulite growth while it is cooled to room temperature. Since you will use PEO powder, you need to try to avoid air bubbles trapped in the melt.

We will monitor the spherulite growth, size and shape, by taking optical images, under crossed polarizers, as a function of time, and compare the spherulite growth rate at these cooling conditions. Let's make a plot of the radius of the spherulite, $R(t)$, as a function of time, t, and discuss the effect of the melt temperature or cooling rate on crystallization.

Polyethylene glycol (PEG) = Poly(ethylene oxide) (PEO)

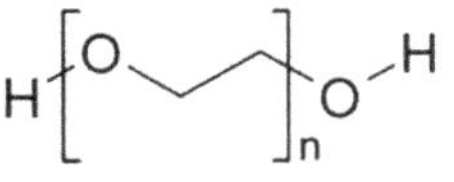

IUPAC namepoly(oxyethylene) (POE) {structure-based}, poly(ethylene oxide) (PEO) {source-based} [1]

Other names Carbowax, GoLYTELY, GlycoLax, Fortrans, TriLyte, Colyte, Halflytely, Macrogol, MiraLAX, MoviPrep

Identifiers

CAS number	25322–68–3 ✓
ChEMBL	CHEMBL1201478 â

Properties

Molecular formula	$C_{2n}H_{4n+2}O_{n+1}$
Molar mass	Variable

Hazards

Flash point	182 °C–287 °C

✗ (verify) (what is: ✓/✗?) Except where noted otherwise, data are given for materials in their standard state (at 25 °C, 100 kPa)

Infobox references

---

http://en.wikipedia.org/wiki/Polyethylene_glycol

References

[1] *Crystallization of polymers.* Available from: http://en.wikipedia.org/wiki/Crystallization_of_polymers

[2] Johnson W A and Mehl R R 1939 Reaction kinetics in processes of nucleation and growth *Trans. Am. Ins. Min. Metall. Eng.* **135** 416–42

[3] Avrami M 1939 Kinetics of phase change. I. General theory *J. Chem. Phys.* **7** 1103–12

[4] Avrami M 1940 Kinetics of phase change. II. Transformation-time relations for distribution of nuclei *J. Chem. Phys.* **8** 212–24

[5] Avrami M 1941 Granulation, phase change, and microstructure: kinetics of phase change. III *J. Chem. Phys.* **9** 177–84

[6] Kolmogorov A 1937 On the statistical theory of metal crystallization *Izv. Akad. Nauk SSSR Seriya. Mat.* **1** 335–60

[7] Barmak K 2018 A commentary on: 'Reaction kinetic in processes of nucleation and growth' *Metall. Mater. Trans.* B **49** 3616–80

[8] Milner S T 2011 Polymer crystal-melt interfaces and nucleation in polyethylene *Soft Matter* **7** 2909–17

[9] Marentette J M and Brown G R 1993 1993 Polymer spherulites. I. Birefringence and morphology *J. Chem. Ed.* **70** 435–9
Marentette J M and Brown G R Polymer spherulites. II. Crystallization kinetics *J. Chem. Ed.* **70** 435–9

[10] Frank C W, Rao V, Despotopoulou M M, Pease R F W, Hinsberg W D, Miller R D and Rabolt J F 1996 Structure in thin and ultrathin spin-cast polymer films *Science* **273** 912–5

IOP Publishing

Through a Modern Microscope
Optical experiments on soft matter
Mohan Srinivasarao, Jung Ok Park and Karthik Nayani

Chapter 8

Experiment 7: Structural color and iridescence of *Morpho* butterfly

8.1 Introduction

The world that we inhabit is incredibly beautiful as well as colorful and has attracted the attention of both scientists and artists alike, perhaps for different reasons. In an effort to understand the world around us starting from the blue color of the sky, the red or pinkish sunsets, the green color of the leaves and the brilliant colors of various insects and birds, humans have been interested in understanding the origins of such colors. In particular, the color of various insects, like beetles and butterflies, in the natural world has attracted the attention of scientific giants such as Hooke, Newton, Michelson, Lord Rayleigh and Sir C V Raman, to name just a few.

Brilliant, iridescent colors found on the bodies and wings of many birds, beetles, butterflies, and moths are produced by light interacting with the structural variations, and hence named *structural colors*, and these have been the subject of study for centuries since at the least the time of Robert Hooke (1635–1703). It should be noted that some structural coloration occurs in combination with pigments. For example, peacock tail feathers are pigmented brown, but it is their microscopic structure that makes them also produce/reflect blue, turquoise and green, and such colors are often iridescent. Iridescence implies that the colors are angle dependent. Robert Hooke was fascinated by the splendid colors found on the tail feathers of a peacock and he asserted they were 'fantastical' as the colors could be made to disappear by wetting the feathers. In his 1665 book, *Micrographia*, Robert Hooke goes on to describe the 'fantastical' colors of the peacock's feathers as follows:

> 'The parts of the Feathers of this glorious Bird appear, through the Microscope, no less gaudy then do the whole Feathers; for, as to the naked eye 'tis evident that the stem or quill of each Feather in the tail sends out multitudes of Lateral branches, … so each of those threads in

doi:10.1088/978-0-7503-1431-2ch8

> the Microscope appears a large long body, consisting of a multitude of bright reflecting parts. … their upper sides seem to me to consist of a multitude of thin plated bodies, which are exceeding thin, and lie very close together, and thereby, like mother of pearl shells, do not onely reflect a very brisk light, but tinge that light in a most curious manner; and by means of various positions, in respect of the light, they reflect back now one colour, and then another, and those most vividly. Now, that these colours are onely fantastical ones, that is, such as arise immediately from the refractions of the light, I found by this, that water wetting these colour'd parts, destroy'd their colours, which seem'd to proceed from the alteration of the reflection and refraction'.

Sir Isaac Newton suggested that such brilliant iridescent colors in birds and insects might be due to the presence of 'thin-film structures', as he had observed such colors while studying the color-producing properties of thin films. Newton, in his '*Treatise on Opticks*' (London, 1704), elaborates on such colors as follows:

> The finely coloured Feathers of some birds, and particularly those of Peacocks Tails, do in the very same part of the Feather appear of several Colours in several positions of the Eye, after the very same manner that thin plates were found to do… and therefore arise from the thinness of the transparent parts of the Feathers; that is, from the slenderness of the very fine Hairs, or *Capillamenta*, which grow out of the sides of the grosser lateral branches or fibers of those Feathers.

As the wave theory of light took root, it became abundantly clear that interference phenomena played a significant role in the color of many insects, beetles and avian feathers. The idea that such colors arise primarily from physical effects such as diffraction or interference as opposed to the colors that are normally produced by the presence of chromophores that absorb or emit light began to take root. Such brilliant colors have been described as metallic colors due to the *saturation or purity* of the colors produced. Since the time of Newton, much progress has been made and it is now recognized that many of these brilliant colors are due to various kinds of microstructures or nanostructures present on butterfly wing scales [1–4]. In other words, the colors are produced by periodic structures of cuticle-air that mimic photonic crystals. Common examples of physical colors are some butterfly wings, hummingbird feathers, arthropod cuticles, the color of Indigo snake skin, gemstones such as opal, and some crystals such as potassium chlorate. The interested reader is referred to an excellent text, *The Physics and Chemistry of Color: The Fifteen Causes of Color* by Kurt Nassau [5].

Many butterflies across all five major families (Nymphalidae, Lycaenidae, Riodinidae, Papilionidae, and Pieridae) display striking colors that result from the nano-morphology on the surface of their wing scales. These structural colors can include deep blacks, reds, oranges and greens as well as the more common blues, violets, ultraviolets, and whites. Typically these colors appear metallic due to the saturation or the purity of the colors produced [1, 5].

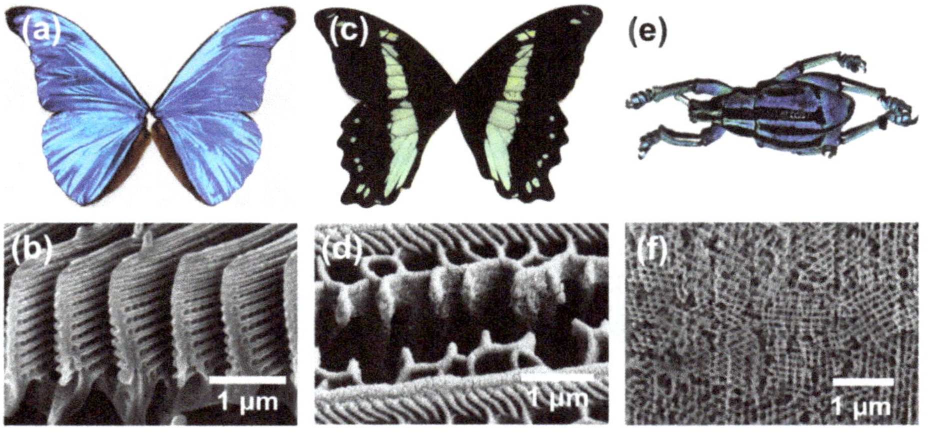

Figure 8.1. (a) *Morpho rhetenor.* (b) Cross-section of the single scale. Each ridge has a tree-shaped multilayer structure. (c) *Papilio nireus.* (d) The cross-sectional view of a single scale prepared by focused ion beam (FIB) milling. It depicts 2D arrangement of inverse cylinders in the *P. nireus.* (e) A weevil, *Eupholus bennetti.* (f) A magnified scanning electron micrograph image of multi-domained 3D crystals exist inside the scale.

The colors perceived by an observer, result from the various types of interaction of the incident light with the nanostructure on the wings of the butterfly scales or the beetle exocuticle or the avian feathers. This interaction may lead to the incident light being scattered in all directions, producing a color that is independent of the viewing angle (as in the feathers of the common bird in North America, the *Bluejay*), or it can have a strong directionality, producing colors that are only seen at certain angles. Physicists, biologists and materials scientists alike have been drawn to study the nano-morphology of these scales, by randomly sampling a few specimens across the 17 000-specimen-rich butterfly superfamily (the Papilionoidae). This initial sampling has uncovered a relatively broad range of morphological structures that are involved in producing the vivid structural colors, and figure 8.1 shows the variations in nanostructures that lead to the production of such colors, in two butterflies and a weevil.

To illustrate that butterfiles use a variety of optical tricks/effects, let us consider as an example, the brilliant green color on the wings of a butterfly, *Papilio palinurus.* The color originates from the hierarchical microstructure of individual wing scales that are tiled on the wing (see figure 8.2). Each wing scale is about 100 μm long and contains multiple 4–10 μm diameter bowls. Each of these bowls contains a multi-layer stack of eleven alternating layers of air and chitin, each about 75 nm thick. Vukusic *et al* [6] elucidated that the distinct green color of the wings is the result of *additive color mixing of yellow and blue*, involving a combination of two reflections off the multilayers in the bowl; the yellow color is due to the reflection from the bottom of the bowl at *normal incidence* and the blue results from two 45° reflections at the edge of the bowl. The blue color is due to the fact that the incident white light sees the multilayer stack present in the bowls at an angle of 45°, and hence the light

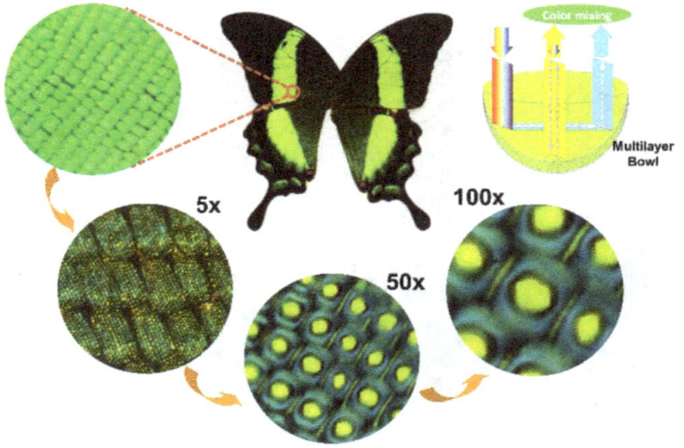

Figure 8.2. (Center) Photograph of *P. palinurus*. (Right) Schematic illustration of ray tracing and color mixing in the individual bowl at the green colored area of the wing. Images of the areas at successively higher magnification (in reflection mode) are optical microscope images (reflection mode) of the bowl structure on the single scale of the wing without polarizers.

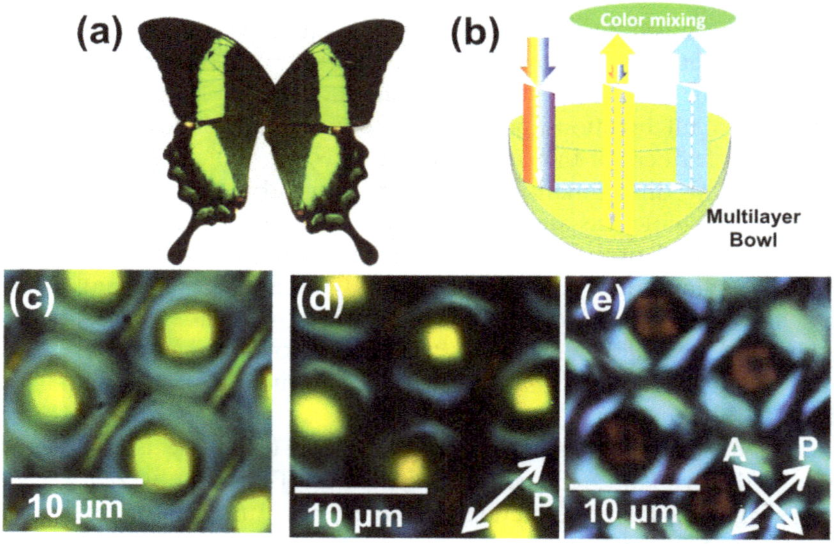

Figure 8.3. (a) Photograph of *P. palinurus*. (b) Schematic description of ray tracing and color mixing in the individual bowl at the green colored area of the wing. Panels (c), (d), and (e) are optical microscope images (reflection mode) of the bowl structure on the single scale of the wing without polarizers (c), with a polarizer in the incident light path (d), and with cross-polarized setup (e), respectively. Adapted from [7] with permission from Springer Nature.

reflected from the multilayer stack is blue-shifted, and thus appears blue. The polarization of the ray reflecting from the edge is changed upon each reflection; hence under crossed polarizers the blue color is not extinguished and is clearly visible under a reflected light microscope. This is illustrated in figure 8.3.

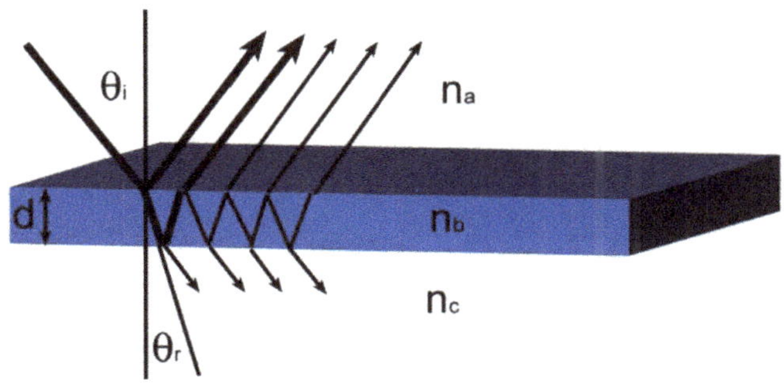

Figure 8.4. Schematic diagram for thin-film interference.

The elegance of such structures is the inspiration of this particular laboratory, focused on structural colors. While we recognize that structural colors are not the only color production mechanism, *we choose to limit our discussion to instances where structural color production is the primary source of color on the wings of butterflies and beetles.*

8.2 Structural color—thin film

The simplest structural color element is a thin film, where the layer of a material is thin enough to be comparable to the wavelength of light. The effect can be easily seen in soap films or oil slicks on water. The thickness of the film and the refractive index of the material determine the optical characteristics of this structure. When a ray of light is incident upon a thin film at an angle θ_i, some of the light is reflected and some is refracted in the film at an angle θ_r. Some of the light is then reflected at the lower film surface, interfering with the light that was originally reflected at the first surface, as illustrated in figure 8.4.

If the film is freestanding, like a soap bubble, the condition for constructive interference is

$$2n_b\, d \cos\theta_r = (m - 1/2)\lambda \tag{8.1}$$

where n_a is the refractive index of the outside medium (air), n_b is the refractive index of the film, d is the thickness of the film, θ_i is the angle of incidence, θ_r is the refracted angle, related to the angle of incidence by Snell's law ($n_b \sin\theta_r = n_a \sin\theta_i$), λ is the wavelength of peak reflectivity and m is an integer. From equation (8.1), we can see that the reflected light is blue-shifted with increasing angle of incidence.

8.3 Structural color—multilayer

A stack of thin films atop each other is called a multilayer. It is perhaps one of the most common structural color features present in Nature. The optical response is governed by multiple reflections from all the layers in the stack. To calculate the reflection, we conveniently package each reflection by Fresnel equation into a matrix form and multiply the matrices in the correct order. The whole procedure is

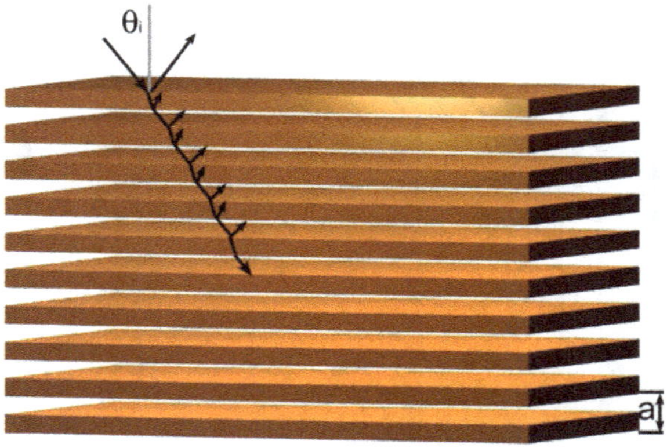

Figure 8.5. Schematic of a multilayer stack.

conveniently available in a free program called 'FreeSnell' [8]. Like with the thin film, the peak reflectance blue shifts with increasing the angle of incidence. The peak width depends on the refractive index contrast between the two materials used and the peak location depends on the thickness of the repeat unit. The repeat unit consists of two layers, one of each material. The maximum reflection is achieved, if the repeat unit a is an integer multiple of $\lambda/2$, making the thickness of each layer $\lambda/4$. Multilayer principles can be a powerful tool for designing optical components, such as dichroic mirrors, dielectric mirrors, interference filters and even an omni-directional reflector.

In butterflies and beetles, the two materials composing the multilayer are often air and chitin. In a sense, only one solid material is used, the other being air. A good example is the butterflies in the *Morpho* family. Their metallic blue color is the result of multilayer reflection, combined with a grating-like structure. The multilayer stack in this case consists of plates supported by a central structure, making the appearance of a tree in the cross-section (see figure 8.5).

8.4 *Morpho rhetenor*

Among the animals that exhibit structural color, the *Morpho* butterflies are perhaps the best known and one of the most spectacular. The bright blue metallic color is produced by tree-like nanostructures on the wing scales. These structures are placed closely together and they interact with light in two ways: (i) as a diffraction grating, resulting from the spacing of the 'trees' and (ii) as a multilayer, resulting from the spacing of the 'branches'. The multilayer is slightly tilted relative to the plane of the scale and thus the strongest reflection is often observed to be about 10° from the surface normal. Usually, the *Morpho* butterflies have two separate sets of scales—transparent cover scales where the tree-like structures are sparse, and colored ground scales, where the tree-like structures are densely packed. In addition, the ground scales have melanin pigment near the bottom of the scale. The particular species *M. rhetenor* is known to lack cover scales and also to have the greatest number of 'branches' on the tree-like structure, as illustrated in figure 8.6.

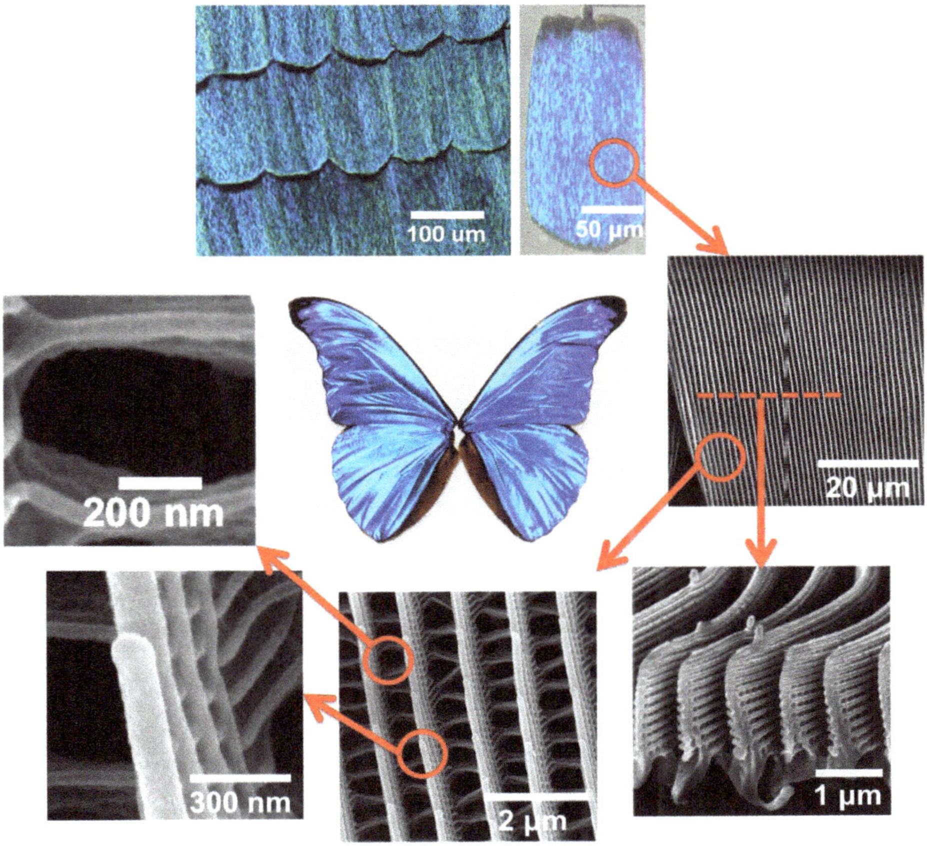

Figure 8.6. Photo of *Morpho rhetenor* in the center. The wing is composed of tiled structure of scales, and an optical microscope image of the single scale detached from the wing of *M. rhetenor*. Scanning electron microscope images, *clockwise*: SEM image of the single scale of *M. rhetenor*, cross-section of the single scale—each ridge has a Fir-tree-shaped multilayer structure. A magnified view of the ridge and cross-rib of the scale, as indicated by the red arrows.

8.5 Experiments

(1) Observe the angular dependent color change of *Morpho* wings, by viewing the wing at various angles.
(2) Observe the wing and the single scale of *M. rhetenor* using an optical microscope under reflection mode. Take images at various magnifications.
(3) Depsoit isopropyl alcohol (IPA) on the wing and observe how the metallic blue color of *M. rhetenor* changes to green. By depositing IPA, the air layer in multilayer structure of single scale will be substituted by IPA and the refractive index difference will be changed. Try to capture the color changes upon the IPA deposition and during the IPA evaporation.

(4) Use multilayer simulator, 'FreeSnell', in order to calculate optical response from multilayer structure of *M. rhetenor*. It is very simple software and we will learn how we put parameters (refractive index, thickness of each stack, and number of stacks) in FreeSnell. Although the multilayer structures in *M. rhetenor* are divided by tree-shaped structures, we will assume simple multilayer structures. Through the calculation, we'll learn how thickness of each layer, number of stacks, refractive index, incident angle, and viewing angle affect resulting coloration of the multilayer structure.

References

[1] Srinivasarao M 1999 Nano-optics in the biological world: beetles, butterflies, birds, and moths *Chem. Rev.* **99** 1935

[2] Kinoshita S, Yoshioka S and Miyazaki J 2008 Physics of structural colors *Rep. Prog. Phys.* **71** 07640

[3] Kinoshita S, Yoshioka S and Kawagoe K 2002 Mechanisms of structural colour in the *Morpho* butterfly: cooperation of regularity and irregularity in an iridescent scale *Proc. R. Soc. Lond.* B **269** 1417–21

[4] Vukusic P, Sambles J R, Lawrence C R and Wootton R J 1999 Quantified interference and diffraction in single Morpho butterfly scales *Proc. R. Soc. Lond.* B **266** 1403–11

[5] Nassau K 2008 *The Physics and Chemistry of Color: The Fifteen Causes of Color* (New York: Wiley)

[6] Vukusic P, Sambles J R and Lawrence C R 2000 Colour mixing in the wingscales of a butterfly *Nature* **404** 457

[7] Srinivasarao M, Iannacchione G S and Parikh A N 2019 Biologically inspired far-from-equilibrium materials *MRS Bull.* **44** 91–95

[8] The FreeSnell thin-film optical simulator, http://people.csail.mit.edu/jaffer/FreeSnell/

www.ingramcontent.com/pod-product-compliance
Lightning Source LLC
LaVergne TN
LVHW062240200926
848484LV00006B/1583

9780750314329